T0326326

COMPANION TO
CHEMICAL THERMODYNAMICS
Basic Theory and Methods

SIXTH EDITION

COMPANION TO CHEMICAL THERMODYNAMICS
Basic Theory and Methods

SIXTH EDITION

IRVING M. KLOTZ
Morrison Professor Emeritus
Northwestern University

ROBERT M. ROSENBERG
MacMillen Professor Emeritus
Lawrence University
Visiting Professor of Chemistry
Northwestern University

A Wiley-Interscience Publication
JOHN WILEY & SONS, INC.
New York / Chichester / Weinheim / Brisbane / Singapore / Toronto

For ordering and customer service, call 1-800-CALL-WILEY.

Library of Congress Cataloging in Publication Data:

ISBN 0-471-37220-X (paper)

PREFACE

We have prepared this solutions manual to assist those who teach and learn from <u>Chemical Thermodynamics.</u> We hope that our choice of style and substance is helpful. We would like to hear from readers about any errors that they find. Graphs in this work have been prepared with PSIPLOT6 from Polysoftware International, and fitting of equations have been done with the same software.

I. M. Klotz
R. M. Rosenberg

Evanston, Illinois

CONTENTS

Chapter 2

Mathematical Preparation for Thermodynamics

2. a) $(1 \text{ cal})/[24.21 \text{ cal } (\text{L atm})^{-1}] = .04131 \text{ L atm}$

    $(.04131 \text{ L atm})(10^3 \text{ cm}^3 \text{ L}^{-1})(10^{-6} \text{ m}^3 \text{ cm}^{-3}) = 4.131 \times 10^{-5} \text{ m}^3 \text{ atm}$

    b) $(1 \text{ cal})(4.184 \text{ J cal}^{-1}) = 4.184 \text{ J} = 4.184 \text{ V C}$

    $(4.184 \text{ V C})/[9.648456 \times 10^4 \text{ C } (\text{Faraday})^{-1}]$

    $= 4.336 \times 10^{-5} \text{ V Faraday}$

4. $$D^2 = H^2 + B^2 \qquad\qquad P = H + B + D \qquad\qquad A = (1/2)BxH$$

    a) $(\partial A/\partial H)_B = B/2$, but we do not know the value of B.

    But we do know the values of H and $(\partial H/\partial B)_D$. Therefore we can use the relationship $H^2 = D^2 - B^2$ to find that $(\partial H/\partial B)_D = -B/H$.

    Then,

    $$B = - H \,(\partial H/\partial B)_D = (-1000 \text{ cm}) \,(-0.5) = 500 \text{ cm}$$

    and $(\partial A/\partial H)_B = 250 \text{ cm}$

    $$(\partial A/\partial B)_D = (\partial A/\partial B)_H + (\partial A/\partial H)_B \,(\partial H/\partial B)_D$$

    $$= 500 \text{ cm} + (250 \text{ cm}) \,(-0.5) = 375 \text{ cm}$$

    $$(\partial A/\partial H)_P = (\partial A/\partial H)_B + (\partial A/\partial B)_H \,(\partial B/\partial H)_P$$

    $$= 250 \text{ cm} + ((500 \text{ cm}) \,(- 1.309) = - 405 \text{ cm}$$

    b) $(\partial H/\partial A)_B = 2/B = 2/(4 \text{ cm}) = 0.5 \text{ cm}^{-1}$

    $(\partial H/\partial B)_A = (-2A)/B^2 = -2.0$

1

$A = B^2 = 16 \text{ cm}^2$

$H = (2A)/B = (32 \text{ cm}^2)/(4 \text{ cm}) = 8 \text{ cm}$

$D = (A^2 + B^2)^{0.5} = [(16 \text{ cm})^2 + (4 \text{ cm})^2]^{0.5} = 16.5 \text{ cm}$

$P = H + B + D = 16 \text{ cm} + 4 \text{ cm} + 16.5 \text{ cm} = 36.5 \text{ cm}$

$(\partial B/\partial A)_P = (\partial B/\partial A)_H + (\partial B/\partial H)_A \, (\partial H/\partial A)_P$

$$= (2/H) - [(2A)/H^2](-0.310 \text{ cm}^{-1}) = 0.41 \text{ cm}^{-1}$$

Since $P = H + B + D$

$(\partial P/\partial A)_B = (\partial H/\partial A)_B + (\partial D/\partial A)_B$

$$= (\partial H/\partial A)_B + (\partial D/\partial H)_B \, (\partial H/\partial A)_B$$

$$= 2/(4 \text{ cm}) + [(8 \text{ cm})/(16.5 \text{ cm})][2/(4 \text{ cm})] = 0.74 \text{ cm}^{-1}$$

6. a) $(\partial H/\partial T)_P = (\partial U/\partial T)_P + P \, (\partial V/\partial T)_P$

If $U = f(T,V)$

$(\partial U/\partial T)_P = (\partial U/\partial T)_V + (\partial U/\partial V)_T \, (\partial V/\partial T)_P$

and

$(\partial H/\partial T)_P = (\partial U/\partial T)_V + [P + (\partial U/\partial V)_T] \, (\partial V/\partial T)_P$

b) $U = H - PV$

$(\partial U/\partial T)_V = (\partial H/\partial T)_V - V \, (\partial P/\partial T)_V$

If $H = f(T,P)$

$(\partial H/\partial T)_V = (\partial H/\partial T)_P + (\partial H/\partial P)_T \, (\partial P/\partial T)_V$

and

$$(\partial U/\partial T)_V = (\partial H/\partial T)_P + [(\partial H/\partial P)_T - V] \, (\partial P/\partial T)_V$$

or

$$(\partial H/\partial T)_P = (\partial U/\partial T)_V + [V - (\partial H/\partial P)_T] \, (\partial P/\partial T)_V$$

c)   From part b)

$$(\partial U/\partial T)_V = (\partial H/\partial T)_P + [(\partial H/\partial P)_T - V] \, (\partial P/\partial T)_V$$

If $H = f(T,P)$ and $dH = 0$

$$0 = (\partial H/\partial T)_P \, (\partial T/\partial P)_H + (\partial H/\partial P)_T$$

and

$$(\partial H/\partial P)_T = - (\partial H/\partial T)_P \, (\partial T/\partial P)_H$$

Thus

$$(\partial U/\partial T)_V = (\partial H/\partial T)_P - [(\partial H/\partial T)_P \, (\partial T/\partial P)_H + V] \, (\partial P/\partial T)_V$$

8. $L = f(T,\tau)$

$$dL = (\partial L/\partial T)_\tau \, dT + (\partial L/\partial \tau)_T \, d\tau$$

At constant L,

$$0 = (\partial L/\partial T)_\tau + (\partial L/\partial \tau)_T \, (\partial \tau/\partial T)_L$$

and

$$(\partial \tau/\partial T)_L = - (\partial L/\partial T)_\tau \, / \, (\partial L/\partial \tau)_T$$

$$= - (\alpha L)/ \, [L/(AY)]$$

$$= - \alpha A Y$$

10.   a) $dL = [L/(YA)] \, d\tau + \alpha L \, dT$

$$[\partial(L/YA)/\partial T]_\tau = [1/(YA)](\partial L/\partial T)_\tau = (\alpha L)/(YA)$$

$$[\partial(\alpha L)/\partial \tau]_T = \alpha(\partial L/\partial \tau)_T = (\alpha L)/(YA)$$

Thus, dL is an exact differential.

b) $dW = \tau dL = (\tau L)/(YA)\ d\tau + \alpha\tau L\ dT$

$[\partial(\tau L/YA)/\partial T]_\tau = (\tau/YA)\ (\partial L/\partial T)_\tau = (\tau\alpha L)/(YA)$

$[\partial(\alpha\tau L)/\partial\tau)_T = \alpha L + \alpha\tau\ (\partial L/\partial T)_\tau = \alpha L + (\tau\alpha L/YA)$

Thus, dW is not an exact differential.

12. $(\partial\kappa/\partial T)_P = -\ (1/V)\ (\partial^2 V)/(\partial T\ \partial P) + (1/V^2)\ (\partial V/\partial T)_P\ (\partial V/\partial P)_T$

$(\partial\beta/\partial P)_T = (1/V)\ [\partial^2 V/(\partial P\ \partial T)] - (1/V^2)\ (\partial V/\partial T)_P\ (\partial V/\partial P)_T$

Since $[\partial^2 V/(\partial T\ \partial P)] = [\partial^2 V/(\partial P\ \partial T]$

$(\partial\beta/\partial P)_T + (\partial\kappa/\partial T)_P = 0$

14. $dG = (\partial G/\partial T)_P\ dT + (\partial G/\partial P)_T\ dP$

$= -\ SdT + VdP$

$(\partial S/\partial P)_T = -\ (\partial^2 G/\partial P\ \partial T)$

and $(\partial V/\partial T)_P = (\partial^2 G/\partial T\ \partial P)$

Since $(\partial^2 G/\partial P\ \partial T) = (\partial^2 G/\partial T\ \partial P)$

$(\partial S/\partial P)_T = -\ (\partial V/\partial T)_P$

16. $dU_m = C_{Vm}\ dT + (a/V_m^2)\ dV_m$

a)
$$\int_1^2 dU = \int_1^2 C_{Vm}dT + \int_1^2 \frac{a}{V_m^2}dV_m$$

dU_m can be integrated to obtain an explicit function for U_m relative to   a

reference value, not an absolute value, if C_{Vm} is known as a function   of T.

b) If dU_m is an exact differential, $(\partial C_{Vm}/\partial V_m)_T = [(\partial/\partial T)(a/V_m^2)] = 0$,

since a is independent of T as well as of V.

Chapter 3

The First Law of Thermodynamics

2.   Let Figure 3-3 in the text be modified as shown in Figure 3-1 below to let F represent the magnitude of the cohesive force of the soap film, and let F' represent the magnitude of the external force that stretches the film. Substitute L in Figure 3-1 for A in Figure 3-3 in the text to represent the length of the wire that defines the moving edge of the soap film. For the system in this problem, the sign in Equation 3-3 in the text conforms with the sign convention we have chosen; that is, DW is positive if ds is positive (upward). negative if ds is negative (downward).

$$DW = F'\,ds$$

$$= F\,ds \quad \text{since the process is reversible}$$

$$= \gamma L\,ds$$

$$= \gamma\,dA \quad \text{where } L ds = dA$$

$$\text{so} \quad W = \int \gamma\,dA$$

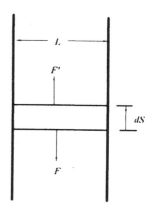

Figure 3-1. Forces on a soap film.

4. $DQ = dU - DW$

If only PV work is done, and the pressure of the system is equal to the constant pressure
of the surroundings,

$$DW = - P_{ex}dV$$

$$= - PdV$$

If P is constant, VdP is equal to zero and can be added without changing the equation. Thus

$$DQ_P = dU + PdV + VdP$$

$$= dU + d(PV)$$

$$= d(U + PV)$$

Since U, P, and V are state functions, $U + PV$ is a state function, and DQ_P is an exact
differential.

Chapter 4

Enthalpy, Enthalpy of Reaction, and Heat Capacity

2.   In the 1982 National Bureau of Standards Tables,

$$\Delta_f H^\circ_m = -277.69 \text{ kJ mol}^{-1}$$

In the International Critical Tables, Vol. V, p. 181, 1928,

$$\Delta_f H^\circ_m = -275.8 \text{ kJ mol}^{-1}$$

The value of the joule changed by about 2 parts per thousand between 1928 and 1982, but not enough to account for the difference in values of $\Delta_f H^\circ_m$. The change in standard state from 1 atm in 1928 to 1 bar in 1982 is also insufficient to explain the difference because ΔH is insensitive to changes in pressure. The value of $\Delta_f H^\circ_m$ is calculated from the enthalpies of combustion of ethanol, carbon (graphite), and hydrogen; the difference in the values for $\Delta_f H_m^\circ$ reflects real changes in the experimental measurements.

The value of $\Delta_f H_m^\circ$ for gaseous ethanol is given in Table 4-5 as -235.0 kJ mol^{-1}. If we subtract the value of $\Delta H_{m,vap}^\circ$, given in the NIST Webbook as 42.26 kJ mol^{-1}, we obtain a value of - 273.3 kJ mol^{-1}, which   is in good agreement with the value from the NBS Tables.

4. $4 \text{ C(graphite)} + 5H_2(g) + S(c) = C_2H_5\text{-}S\text{-}C_2H_5(g)$ (1)

$$\Delta H_{m1}^\circ = -147.24 \text{ kJ mol}^{-1}$$

$4 \text{ C(graphite)} + 5H_2(g) + 2S(c) = C_2H_5\text{-}S\text{-}S\text{-}C_2H_5(g)$ (2)

$$\Delta H_{m2}^\circ = -201.92 \text{ kJ mol}^{-1}$$

$$S(c) = S(g) (3)$$

$$\Delta H_{m3}^\circ = 278.805 \text{ kJ mol}^{-1}$$

Reaction (1) - Reaction (2) + Reaction (3) yields

C_2H_5-S-S-C_2H_5(g) = C_2H_5-S-C_2H_5(g) + S(g)

and $\Delta H_m^\circ = \Delta H_{m1}^\circ - \Delta H_{m2}^\circ + \Delta H_{m3}^\circ$

$= -147.24 \text{ kJ mol}^{-1} + 201.92 \text{ kJ mol}^{-1} + 278.805 \text{ kJ mol}^{-1}$

$= 333.49 \text{ kJ mol}^{-1} = \varepsilon_{S\text{-}S}$

because the reaction involves only the breaking of one S-S bond in the gas phase. Although one

C-S bond is broken in the reaction, another is formed.

6. $\qquad N_2H_4(g) = N_2H_2(g) + H_2(g) \qquad \Delta H_m^\circ = 109 \text{ kJ mol}^{-1}$

$N_2(g) + 2H_2(g) = N_2H_4(g) \qquad \Delta H_m^\circ = 95.0 \text{ kJ mol}^{-1}$

The sum of these two reactions is

$N_2(g) + H_2(g) = N_2H_2(g) \qquad \Delta H_m^\circ = 204 \text{ kJ mol}^{-1}$

$\Delta H_m^\circ = \varepsilon_{N\equiv N} + \varepsilon_{H\text{-}H} - \varepsilon_{N=N} - 2\varepsilon_{N\text{-}H}$

Thus, $\varepsilon_{N=N} = \varepsilon_{N\equiv N} + \varepsilon_{H\text{-}H} - 2\varepsilon_{N\text{-}H} - 204 \text{ kJ mol}^{-1}$

$= 944.7 \text{ kJ mol}^{-1} + 435.89 \text{ kJ mol}^{-1} - 2(390.8 \text{ kJ mol}^{-1})$

$-204 \text{ kJ mol}^{-1} = 395 \text{ kJ mol}^{-1}$

8. The bond enthalpy of the I-Cl bond is the value of ΔH_m° for the reaction:

ICl(g) = I(g) + Cl(g)

In terms of standard enthalpies of formation,

$\Delta H_m^\circ = \Delta_f H_m^\circ[\text{I(g)}] + \Delta_f H_m^\circ[\text{Cl(g)}] - \Delta_f H_m^\circ[\text{ICl(g)}]$

$= 106.838 \text{ kJ mol}^{-1} + 121.679 \text{ kJ mol}^{-1} - 17.78 \text{ kJ mol}^{-1}$

$= 210.74 \text{ kJ mol}^{-1}$

Table 4-3 gives a value of 209 kJ mol^{-1}, which is good agreement.

10. a) $\qquad S_8(g) = 8S(g)$

$$\Delta H_m{}^\circ = 8\Delta_f H_m{}^\circ[S(g)] - \Delta_f H_m{}^\circ[S_8(g)]$$

$$= 8(278.805 \text{ kJ mol}^{-1} - 102.30 \text{ kJ mol}^{-1}$$

$$= 2128.14 \text{ kJ mol}^{-1} = 8\varepsilon_{S-S}$$

$$\varepsilon_{S-S} = 266.02 \text{ kJ mol}^{-1}$$

b) $\qquad$ $S_2Cl_2(g) = 2S(g) + 2Cl(g)$

$$\Delta H_m{}^\circ = 2\Delta_f H_m{}^\circ[S(g)] + 2\Delta_f H_m{}^\circ[Cl(g)] - \Delta_f H_m{}^\circ[S_2Cl_2]$$

$$= 2(278.805 \text{ kJ mol}^{-1}) + 2(121.679 \text{ kJ mol}^{-1}) -(- 18.4 \text{ kJ mol}^{-1})$$

$$= 819.4 \text{ kJ mol}^{-1} = \varepsilon_{S-S} + 2\varepsilon_{S-Cl}$$

$$\varepsilon_{S-Cl} = 276.7 \text{ kJ mol}^{-1}$$

c) To find $\Delta_f H_m{}^\circ$ for $SCL_2(g)$, we need to find reactions that sum to

$\qquad$ $S(c) + Cl_2(g) = SCl_2(g)$, such as

$\qquad\qquad$ $S(c) = S(g)$ $\qquad\qquad\qquad\qquad$ (1)

$$\Delta H_{m1}{}^\circ = \Delta_f H_m{}^\circ[S(g)] = 278.805 \text{ kJ mol}^{-1}$$

$\qquad\qquad$ $Cl_2(g) = 2\ Cl(g)$ $\qquad\qquad\qquad$ (2)

$$\Delta H_{m2}{}^\circ = 2\Delta_f H_m{}^\circ[Cl(g)] = 2(121.679 \text{ kJ mol}^{-1}) = 243.358 \text{ kJ mol}^{-1}$$

$\qquad\qquad$ $S(g) + 2Cl(g) = SCl_2(g)$ $\qquad\qquad$ (3)

$$\Delta H_{m3}{}^\circ = -2\varepsilon_{S-Cl} = - 2(276.7 \text{ kJ mol}^{-1})$$

The sum of the three reactions is the desired reaction; therefore

$$\Delta_f H_m{}^\circ[Scl_2(g)] = \Delta H_{m1}{}^\circ + \Delta H_{m2}{}^\circ + \Delta H_{m3}{}^\circ$$

$$= 278.805 \text{ kJ mol}^{-1} + 243.358 \text{ kJ mol}^{-1} - 553.4 \text{ kJ mol}^{-1}$$

$$= - 31.2 \text{ kJ mol}^{-1}$$

All data are from the 1982 NBS Tables of Thermodynamic Properties.

12.   a) $U = H - PV$

Therefore,

$(\partial U/\partial V)_P = (\partial H/\partial V)_P - P$

$= (\partial H/\partial T)_P \, (\partial T/\partial V)_P - P = C_P(\partial T/\partial V)_P - P$

b) $dU = (\partial U/\partial V)_T dV + (\partial U/\partial T)_V dT$

At constant V, $dV = 0$, and

$(\partial U/\partial P)_V = (\partial U/\partial T)_V \, (\partial T/\partial P)_V$

$= C_V \, (\partial T/\partial P)_V$

14.   Water, at 4°C, the temperature of maximum density. At that temperature $(\partial V/\partial T)_P$ is equal to zero.

16.   a) We need to calculate $\Delta H_m°$ for the process

0.5 Ag(s,31°C) + 0.5 Cd(s,31°C) = $Ag_{0.5}Cd_{0.5}$(s,31°C)

The following equations sum to the desired equation:

0.5 Ag(s,31°C) = 0.5 Ag(in liquid Sn,250°C)   (1)

$\Delta H_{m1}° = 10209$ J mol^{-1}

0.5 Cd(s,31°C) = 0.5 Cd(in liquid Sn,250°C)   (2)

$\Delta H_{m2}° = 9498$ J mol^{-1}

0.5 Ag(in liquid Sn,250°C) + 0.5 Cd(in liquid Sn,250°C)

$= Ag_{0.5}Cd_{0.5}$(s,31°C)

$\Delta H_{m3}° = -27447$ J mol^{-1}

$\Delta_f H_m° = \Delta H_{m1}° + \Delta H_{m2}° + \Delta H_{m3}° = -7740$ J mol^{-1}

b)We need to calculate $\Delta H_m°$ for the process

$$0.412 \text{ Ag(s,31°C)} + 0.588 \text{ Cd(s,31°C)} = \text{Ag}_{0.412}\text{Cd}_{0.588}\text{(s,31°C)}$$

The following equations sum to the desired equation:

$$0.412 \text{ Ag(s,31°C)} = 0.412 \text{ Ag(in liquid Sn,250°C)} \qquad (4)$$

$$\Delta H_{m4}° = (0.412)(20418 \text{ J mol}^{-1}) = 8412 \text{ J mol}^{-1}$$

$$0.588 \text{ Cd(s,31°C)} = 0.588 \text{ Cd(in liquid Sn,250°C)} \qquad (5)$$

$$\Delta H_{m5}° = (0.588)(18955 \text{ J mol}^{-1}) = 11169 \text{ J mol}^{-1}$$

$$0.412 \text{ Ag(in liquid Sn,250°C)} + 0.588 \text{ Cd(in liquid Sn,250°C)}$$

$$= \text{Ag}_{0.412}\text{Cd}_{0.588}\text{(s,31°C)} \qquad (6)$$

$$\Delta H_{m6}° = -27949 \text{ J mol}^{-1}$$

$$\Delta_f H_m° = \Delta H_{m4}° + \Delta H_{m5}° + \Delta H_{m6}° = -8368 \text{ J mol}^{-1}$$

18. $NH_4Cl(s) = NH_4^+(g) + Cl^-(g)$ $\qquad \Delta H_m° = 640 \text{ kJ mol}^{-1}$

$NH_4^+(g) = NH_3(g) + H^+(g)$ $\qquad \Delta H_m° = \mathcal{P}_{NH_3}$

$H^+(g) + e^-(g) = H(g)$ $\qquad \Delta H_m° = -1305 \text{ kJ mol}^{-1}$

$Cl^-(g) = Cl(g) + e^-$ $\qquad \Delta H_m° = 387 \text{ kJ mol}^{-1}$

$H(g) = 1/2 \text{ H}_2(g)$ $\qquad \Delta H_m° = -217.965 \text{ kJ mol}^{-1}$

$Cl(g) = 1/2 \text{ Cl}_2(g)$ $\qquad \Delta H_m° = -121.679 \text{ kJ mol}^{-1}$

$NH_3(g) = 1/2 \text{ N}_2(g) + 3/2 \text{ H}_2(g)$ $\quad \Delta H_m° = 45.6 \text{ kJ mol}^{-1}$

$1/2 \text{ N}_2(g) + 2\text{H}_2(g) + 1/2 \text{ Cl}_2(g) = NH_4Cl(s)$ $\qquad \Delta H_m° = -314.2 \text{ kJ mol}^{-1}$

Since the listed equations constitute a thermodynamic cycle, e.g. they sum to zero, the sum of

the $\Delta H_m°$'s is also equal to zero. Thus

$$\mathcal{P} = (-640 + 1305 - 387 + 217.765 + 121.679 + -45.6 + 314.2)\text{kJ mol}^{-1}$$

$$= 886 \text{ kJ mol}^{-1}$$

20. $$H_2(g) = H^+(g) + H(G) + e^-(g) \qquad \Delta H_{m1}^\circ = 18.0 \text{ eV}$$

$$H^+(g) + e^-(g) = H(g) \qquad \Delta H_{m2}^\circ = -13.6 \text{ eV}$$

The sum of reactions (1) and (2) is

$$H_2(g) = 2 \text{ H}(g) \qquad \Delta H_m^\circ = \Delta H_{m1}^\circ + \Delta H_{m2}^\circ = 4.4 \text{ eV}$$

$$= \varepsilon_{H\text{-}H}$$

$$\varepsilon_{H\text{-}H} = (4.4 \text{ eV})[96.44 \text{ kJ mol}^{-1} (\text{eV})^{-1}] = 420 \text{ kJ mol}^{-1}$$

22. $$UO_2(s) + 3F_2(g) = UF_6(s) + O_2(g)$$

$$\Delta H_m^\circ = \Delta_f H_m^\circ[UF_6(s)] - \Delta_f H_m^\circ[UO_2(s)] = -1112.6 \text{ kJ mol}^{-1}$$

Therefore,

$$\Delta_f H_m^\circ[UF_6(s)] = -1085.1 \text{ kJ mol}^{-1}$$

24. $$HC\equiv C\bullet(g) + e^-(g) = HC\equiv C:^-(g) \qquad \Delta H_m^\circ = -2.969 \text{ eV} = -68.43 \text{ kcal mol}^{-1}$$

$$HC\equiv CH:^-(g) + H^+(g) = HC\equiv CH(g) \qquad \Delta H_m^\circ = -377.8 \text{ kcal mol}^{-1}$$

$$H\bullet(g) = H^+(g) + e^- \quad \Delta H_m^\circ = 13.595 \text{ eV} = 313.4 \text{ kcal mol}^{-1}$$

The sum of these three reactions is

$$HC\equiv C\bullet(g) + H\bullet(g) = HC\equiv CH(g) \qquad \Delta H_m^\circ = -132.8 \text{ kcal mol}^{-1}$$

so that ΔH_m° for the dissociation of the C-H bond in acetylene is

equal to 132.8 kcal mol^{-1}.

26.   a) We need to find ΔH_m° for the reaction

$$Cp_2SmI = Cp_2Sm + I \quad \text{(in toluene)} \qquad (1)$$

We know ΔH_m° for the reaction

$$2 Cp_2Sm + I_2 = 2 Cp_2SmI \quad \text{(in toluene)} \qquad (2)$$

$$\Delta H_{m2}{}^\circ = - 102.4 \text{ kcal mol}^{-1}$$

For the reaction

$$2I = I_2 \tag{3}$$

$$\Delta H_{m3}{}^\circ \approx - \varepsilon_{I-I} = - 150.88 \text{ kJ mol}^{-1} = - 36.06 \text{ kcal mol}^{-1}$$

Reaction 2 + Reaction 3 = - 1/2 Reaction 1. Therefore

$$\Delta H_{m1}{}^\circ = - 2 (\Delta H_{m2}{}^\circ + \Delta H_{m3}{}^\circ)$$

$$= - 2 (-102.4 \text{ kcal mol}^{-1} - 36.06 \text{ kcal mol}^{-1})$$

$$= 276.9 \text{ kcal mol}^{-1}$$

b) For the reaction

$$2 Cp_2SmH + I_2 = 2 Cp_2SmI + H_2 \quad \text{(in toluene0} \tag{4}$$

$$\Delta_f H_{m4}{}^\circ = - 104.8 \text{ kcal mol}^{-1}$$

We need to find $\Delta_f H_m{}^\circ$ for the reaction

$$Cp_2SmH = Cp_2Sm + H \quad \text{(in toluene)} \tag{5}$$

If we subtract Equation 2 from Equation 4 and add

$$H_2 = 2H \quad \text{for wich } \Delta_f H_m{}^\circ = \varepsilon_{H-H}$$

then the result is 2 times Equation 5. Thus,

$$\Delta_f H_{m5}{}^\circ = 1/2 (\Delta_f H_{m4}{}^\circ - \Delta_f H_{m2}{}^\circ + \varepsilon_{H-H})$$

$$= 1/2 [-104.8 \text{ kcal mol}^{-1} - (- 102 \text{ kcal mol}^{-1}) + 104 \text{ kcal mol}^{-1}]$$

$$= 51 \text{ kcal mol}^{-1}$$

Chapter 5

Application of the First Law to Gases

2.
$$dW = -P'dV$$

$$W = \int_{V_1}^{V_2} - P'dV = -P'(V_2 - V_1)$$

4. $PV_m = RT + BP$; $V_m = (RT/P) + B$

$$dV_m = -(RT/P^2)dP$$

$$W = -\int_{V_{m1}}^{V_{m2}} P dV_m$$

$$= \int_{P_1}^{P_2} \frac{RT}{P} dP$$

$$= RT \ln \frac{P_2}{P_1}$$

$$P = \frac{RT}{V_m - b}$$

$$\left(\frac{\partial P}{\partial T}\right)_V = \frac{R}{V_m - b}$$

$$\left(\frac{\partial U}{\partial V}\right)_T = T\left(\frac{\partial P}{\partial T}\right)_V - P$$

$$= \frac{RT}{V_m - b} - P = 0$$

Thus $\Delta U = 0$ for an isothermal change for this gas, and

$$Q = -W = -RT \ln \frac{P_2}{P_1}$$

$$\Delta H = \Delta U + \Delta (PV) = \Delta (PV)$$

$$= \Delta (RT + BP) = B(P_2 - P_1)$$

6.

$$C_P = C_V + \left[V - \left(\frac{\partial H}{\partial P} \right)_T \right] \left(\frac{\partial P}{\partial T} \right)_V$$

from Equation 5-67,

$$\mu_{JT} = - \frac{1}{C_P} \left(\frac{\partial H}{\partial P} \right)_T$$

Therefore,

$$C_P = C_V + \left[V + C_P \mu_{JT} \right] \left(\frac{\partial P}{\partial T} \right)_V$$

$$C_V = C_P \left[1 - \mu_{JT} \left(\frac{\partial P}{\partial T} \right)_V \right] - V \left(\frac{\partial P}{\partial T} \right)_V$$

b) At the inversion temperature, $\mu_{JT} = 0$, and

$$C_V = C_P - V \left(\frac{\partial P}{\partial T} \right)_V$$

8.

$$Q = nRT \ln \frac{V_2}{V_1}$$

$$n = \frac{Q}{RT \ln \dfrac{V_2}{V_1}}$$

$$= \frac{9410 \text{J}}{(8.3145 \text{J} \, \text{mol}^{-1} \text{K}^{-1})(298.15 K) \ln \left[(10 \text{dm}^3)/(1.5 \text{dm}^3) \right]} = 2.00 \text{mol}$$

10.   Again, we start with the total differential dU,

$$dU = \left(\frac{\partial U}{\partial T}\right)_V dT + \left(\frac{\partial U}{\partial V}\right)_T dV$$

$$= C_V dT + \left(\frac{\partial U}{\partial V}\right)_T dV$$

Since dU is an exact differential,

$$\left(\frac{\partial C_V}{\partial V}\right)_T = \left(\frac{\partial}{\partial T}\left(\frac{\partial U}{\partial V}\right)_T\right)_V$$

$$= \left(\frac{\partial}{\partial T}(0)\right)_V = 0$$

By an analogous argument, which the student should work through, $\left(\frac{\partial C_P}{\partial P}\right)_T = 0$.

12.

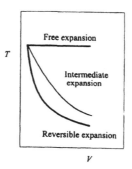

Figure 5-1. A graph of T against V for free, reversible, and irreversible expansions of an ideal gas. As we showed in Table 5-2, the final temperature is the same as the initial temperature in a free, adiabatic expansion.

For the reversible, adiabatic expansion of a monatomic ideal gas, $C_{Vm} = (3R)/2$, so that

$$\frac{T_2}{T_1} = \left[\frac{V_1}{V_2}\right]^{\frac{R}{C_{Vm}}} = \left[\frac{V_1}{V_2}\right]^{\frac{R}{3R/2}} = \left[\frac{V_1}{V_2}\right]^{2/3}$$

so that

$$TV^{2/3} = \text{a constant}$$

whose value depends on the initial temperature, pressure, and number of moles. For the intermediate, irreversible, adiabatic expansion, the work done is less than for the reversible expansion because the constant external pressure is always less than the pressure of the gas. Since

$$W = \int_{T_1}^{T_2} C_{V_m} dt$$

T_2 will be closer to T_1 in the irreversible expansion than in the reversible expansion.

14.

$$P = \frac{RT}{V_m - b} - \frac{a}{V_m^2}$$

If we differentiate with respect to T at constant P, we obtain

$$0 = \frac{(V_M - b)R - RT\left(\dfrac{\partial V_m}{\partial T}\right)_P}{(V_m - b)^2} + \frac{2a}{V_m^3}\left(\frac{\partial V_m}{\partial T}\right)_P$$

$$\left(\frac{\partial V_m}{\partial T}\right)_P = \frac{R(V_m - b)}{RT - (V_m - b)^2 \dfrac{2a}{V_m^3}}$$

$$\beta = \frac{1}{V_m}\left(\frac{\partial V_m}{\partial T}\right)_P = \frac{R\dfrac{V_m - B}{V_m}}{RT - \dfrac{2a}{V_m^3}(V_m - b)^2}$$

16.   The Redlich-Kwong equation is

$$\left[P + \frac{a}{T^{1/2} V_m (V_m + b)} \right] (V_m - b) = RT$$

If we solve for P, the result is

$$P = \frac{RT}{V_m - b} - \frac{a}{T^{1/2} V_m (V_m + b)}$$

and the compressibility Z is

$$Z = \frac{P V_m}{RT} = \frac{V_m}{V_m - b} - \frac{a}{RT^{3/2} (V_m + b)}$$

If we divide numerator and denominator on the right by V_m, we obtain an equation for Z as a function of $(1/V_m)$.

$$Z = \frac{1}{1 - b \dfrac{1}{V_m}} - \frac{a \dfrac{1}{V_m}}{RT^{3/2} \left(1 + b \dfrac{1}{V_m} \right)}$$

The virial equation expresses Z as a polynomial in $(1/V_m)$

$$Z = 1 + B \left(\frac{1}{V_m} \right) + C \left(\frac{1}{v_m} \right)^2 + \cdots$$

If we expand the equation for Z in a McLaurin series about $1/V_m = 0$,

$$B = \left[\frac{\partial Z}{\partial \left(\dfrac{1}{V_m} \right)} \right]_{T, \, 1/V_m = 0} = b - \frac{a}{RT^{3/2}}$$

Chapter 6

The Second Law of Thermodynamics

2.   The Carnot cycles sketched here are accurate plots with high and low temperatures of 600 K and 300 K. The initial and final pressures of the isothermal expansion step are 100 kPa and 10 kPa. The scales for S, U, and H have an arbitrary zero. The absolute values should not be of concern to students; what is important is that you understand the direction of change of each variable in each step.

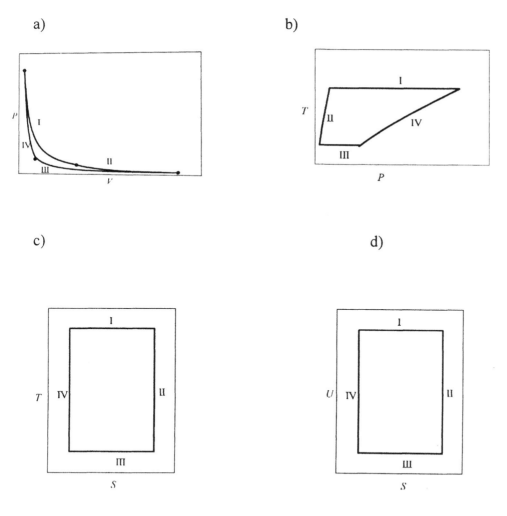

Figure 6-1. Carnot cycles for an ideal gas.

e)   f)

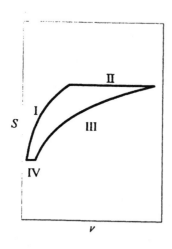

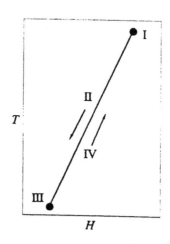

Figure 6-1. (continued)

4.   a) From Eq. 6-130,

$$\left(\frac{\partial U}{\partial V}\right)_T = T\left(\frac{\partial P}{\partial T}\right)_V$$

From the equation of state,

$$P = \frac{RT}{V_m - B}$$

$$\left(\frac{\partial P}{\partial T}\right)_V = \frac{R}{V_m - B} \qquad \text{and}$$

$$\left(\frac{\partial U}{\partial V}\right)_T = T\frac{R}{V_m - B} - P = 0$$

Therefore, U is a function of T only.

b)

$$dU = \left(\frac{\partial U}{\partial T}\right)_V dT + \left(\frac{\partial U}{\partial V}\right)_T dV$$

$$= C_V dT + [T\left(\frac{\partial P}{\partial T}\right)_V - P]dV$$

$$\left(\frac{\partial U}{\partial V}\right)_P = C_V\left(\frac{\partial T}{\partial V}\right)_P + T\left(\frac{\partial P}{\partial T}\right)_V - P$$

Therefore,

$$\left(\frac{\partial U_m}{\partial V_m}\right)_P = C_{Vm}\left(\frac{\partial T}{\partial V_m}\right)_P + T\left(\frac{\partial P}{\partial T}\right)_V - P$$

$$= c_{Vm}\frac{P}{R} + \frac{RT}{V_m - b} - P$$

$$= C_{Vm}\frac{P}{R}$$

For an ideal gas, from the same expression, $(\partial U_m/\partial V_m)_P$ also equals $(C_{Vm} P)/R$.

c) From Eq. 6-109

$$dS_m = \frac{dU_m}{T} + \frac{PdV_m}{T}$$

$$= \frac{C_{Vm}dT}{T} + \frac{RdV_m}{V_m - b}$$

If C_{Vm} is constant, this equation can be integrated to obtain

$$S_m - S_{m0} = C_{Vm} \ln T + R \ln (V_m - b)$$

6. Q_1 and Q_2 must be opposite in sign, and Q_2 must be larger in magnitude than Q_1. Thus, only

octants 3 and 7 contain possible Carnot cycles.

8.   The following sketches indicate the direction of change for each variable in each step.

a)   b)

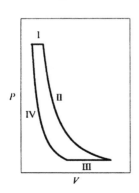   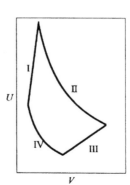

c)   d)

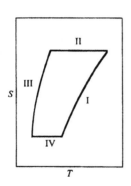   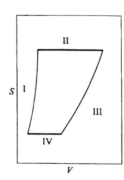

e)   f)

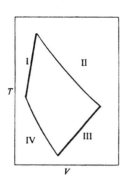   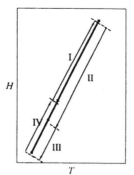

Figure 6-2. Joule cycles for a gas with the equation of state $P(V_m - B) = RT$.

10.   a)

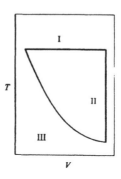

Figure 6-3. A three step cycle for an ideal gas.

b)

$$\left(\frac{\partial S}{\partial V}\right)_T = \left(\frac{\partial P}{\partial T}\right)_V = \frac{R}{V_m - b}$$

$$\Delta S_I = \int_{V_1}^{V_2} \frac{R}{V_m - b} dV = R \ln \frac{V_{m2} - b}{V_{m1} - b}$$

$$\Delta S_{II} = \int_{T_2}^{T_1} \frac{C_V dT}{T} = C_V \ln \frac{T_1}{T_2}$$

$$\Delta S_{III} = 0$$

$$\oint dS = \Delta S_I + \Delta S_{II} + \Delta S_{III}$$

$$= R \ln \frac{V_{m2} - b}{V_{m2} - b} + C_V \ln \frac{T_1}{T_2}$$

In the adiabatic step, $DQ = 0$, and $dU = dW$, so that

$$C_V dT = -PdV = \frac{-RT}{V_m - b} dV$$

$$\frac{C_V \, dT}{T} = \frac{-R \, dV}{V_m - b}$$

$$C_V \ln \frac{T_2}{T_1} = R \ln \frac{V_{m1} - b}{V_{m2} - b}, \text{ so } \oint dS = 0$$

12. If the line ab lies to the left of line ab', then we would have the cycle $a{\to}b{\to}b'{\to}a$. In step I $(a{\to}b)$, $Q_I = 0$, since the step is adiabatic. In step II $(b{\to}b')$, with b to the left of b', $\Delta S_{II} > 0$, and $Q_{II} = T\Delta S_{II} > 0$. In step III $(b'{\to}a)$, $\Delta S = 0$, and $Q_{III} = 0$. Thus, $\oint DQ = Q_{II} > 0$. Since $\oint dU = 0$, $\oint DW = -Q_{II} < 0$. Thus, the cycle has converted heat completely into work without any other change in the system or surroundings, contradicting the Kelvin-Planck statement of the second law of thermodynamics.

14. The second law states that $\Delta S > 0$ for system plus surroundings in any natural process. Although biological systems may decrease the entropy of the system in the process of growth, development, and the synthesis of complex macromolecules, the systems must produce an increase in entropy in the surroundings that is larger in magnitude than the decrease in entropy of the system. This increase comes about by converting large food molecules into CO_2 and H_2O and by giving off heat to the surroundings.

16.

$$\Delta S_I = \frac{Q_I}{T} = \frac{39.3 \text{ kJ}}{400 \text{ K}} = 98.3 \text{ J K}^{-1}$$

$$\Delta S_{II} = \Delta S_{IV} = 0$$

Since S is a state function, $\Delta S_{III} = -\Delta S_I = -98.3 \text{ J K}^{-1}$, and

$$Q_{III} = T\Delta S_{III} = (300 \text{ K})(-98.3 \text{ J K}^{-1}) = -29.4 \text{ kJ}$$

Moles H_2O condensed = (29.4 kJ)/(43.5 kJ mol^{-1}) = 0.68 mol

Mass H_2O condensed = (0.68 mol)(18.02 g mol^{-1}) = 1.22 g

18.   From the footnote on p. 107 in the text, $\gamma = Q_2'/W' = 1/\varepsilon$

$\varepsilon = (0.40)(T_2 - T_1)/T_2 = (0.40)(313 \text{ K} - 277 \text{ K})/(313 \text{ K}) = 0.0460$

$\gamma = 1/\varepsilon = 21.74$

$W' = Q_2'/\gamma = (10^6 \text{ kwh})/(21.74) = 4.60 \times 10^4 \text{ kwh}$

The cost of direct electric heating would be

$(1 \times 10^6 \text{ kwh})(\$0.05/\text{kwh}) = \$50,000$

The cost of heating with the heat pump would be

$(4 \times 10^4 \text{ kwh})(\$0.05/\text{kwh}) = \$2,300$

The annual savings would be \$47,700

In the first year, \$12,000 would be spent in interest on a loan of \$200,000, and \$35,700 could

be applied to principal. In each succeeding year, the interest would be less, and more could be

applied to principal. The loan would be repaid in less than 5 years from the savings due to the

heat pump. Nice if you live on the shore of a large lake. The actual savings would, of course,

be smaller than calculated here.

Chapter 7

Equilibrium and Spontaneity for Systems at Constant Temperature: The Gibbs, Helmholtz, Planck, and Massieu Functions

2.   a) $\Delta G = \Delta H - T\Delta S$

     $\Delta G/T = \Delta H/T - \Delta S$

$$\left[\frac{\partial (\Delta G/T)}{\partial T} \right]_P = \frac{1}{T} \left(\frac{\partial \Delta H}{\partial T} \right)_P - \frac{\Delta H}{T^2} - \left(\frac{\partial \Delta S}{\partial T} \right)_P$$

$$= \frac{C_P}{T} - \frac{\Delta H}{T^2} - \frac{C_P}{T} = - \frac{\Delta H}{T^2}$$

$$\left[\frac{\partial (\Delta G/T)}{\partial (1/T)} \right]_P = \left[\frac{\partial (\Delta G/T)}{\partial T} \right]_P \frac{\partial T}{\partial (1/T)} = \left[\frac{\partial (\Delta G/T)}{\partial T} \right]_P \div \left[\frac{\partial (1/T)}{\partial T} \right]$$

$$= - \frac{\Delta H}{T^2} \div \left(- \frac{1}{T^2} \right) = \Delta H$$

Since $\Delta Y = - \Delta G/T$

$$\left(\frac{\partial \Delta Y}{\partial (1/T)} \right)_P = - \Delta H$$

b) When G is considered as a function of T and P,

$$dG = \left(\frac{\partial G}{\partial T} \right)_P dT + \left(\frac{\partial G}{\partial P} \right)_T dP$$

$$\left(\frac{\partial G}{\partial V} \right)_T = \left(\frac{\partial G}{\partial P} \right)_T \left(\frac{\partial P}{\partial V} \right)_T = V \left(\frac{\partial P}{\partial V} \right)_T$$

Also,

$$dG = \left(\frac{\partial G}{\partial T}\right)_P dT + \left(\frac{\partial G}{\partial P}\right)_T dP$$

$$\left(\frac{\partial G}{\partial T}\right)_V = \left(\frac{\partial G}{\partial P}\right)_T \left(\frac{\partial P}{\partial T}\right)_V + \left(\frac{\partial G}{\partial T}\right)_P$$

$$= V\left(\frac{\partial P}{\partial T}\right)_V - S$$

Since, when G is considered a function of T and V,

$$dG = \left(\frac{\partial G}{\partial V}\right)_T dV + \left(\frac{\partial G}{\partial T}\right)_V dT$$

$$= V\left(\frac{\partial P}{\partial V}\right)_T dV + \left[V\left(\frac{\partial P}{\partial T}\right)_V - S\right] dT$$

4.   a) $dG = dH - TdS - SdT$

$= DU + PdV + VdP - TdS - SdT$

$= DQ + DW + PdV + VdP - TdS - SdT$

If the rubber band is stretched reversibly with no change in volume at constant pressure,

$DQ = TdS$, $PdV = 0$, $VdP = 0$, and $DW = \tau dL$, so that

$dG = \tau dL - SdT$

b)   If we consider G as a function of T and L,

$$dG = \left(\frac{\partial G}{\partial L}\right)_T dL + \left(\frac{\partial G}{\partial T}\right)_L dT$$

Thus,

$$\left(\frac{\partial G}{\partial L}\right)_T = \tau \quad \text{and} \quad \left(\frac{\partial G}{\partial T}\right)_L = -S$$

c)   By the equality of cross derivatives,

$$\left(\frac{\partial \tau}{\partial T}\right)_L = -\left(\frac{\partial S}{\partial L}\right)_T$$

d) $dU = DQ + DW$, and for a reversible change, $DQ = TdS$, so that

$dU = TdS + \tau dL$. Then,

$$\left(\frac{dU}{dL}\right)_T = T\left(\frac{\partial S}{\partial L}\right)_T + \tau = \tau - T\left(\frac{\partial \tau}{\partial T}\right)_L$$

e)   By analogy with an ideal gas, an "ideal" rubber band would be one for which $(\partial U/\partial L)T = 0$. Then

$$\tau = -T\left(\frac{\partial \tau}{\partial T}\right)_L \quad \text{and} \quad \frac{1}{T} = \frac{1}{\tau}\left(\frac{\partial \tau}{\partial T}\right)_L$$

where τ is analogous to P and L is analogous to V.

6.

$$W = -P(V_{mf} - V_{mi})$$

$$= -P\left(0 - \frac{RT}{P}\right) = RT = 3.10 \text{ kJ mol}^{-1}$$

if we neglect the volume of the liquid

$Q = \Delta H_{vap}$, if only PV work is done

$= (18.02 \text{ g mol-1})(-2256.8 \text{ J g-1}) = -40.67 \text{ kJ mol-1}$

$\Delta U = Q + W = -37.57 \text{ kJ mol-1}$

$\Delta H = -40.67 \text{ kJ mol-1}$

$\Delta S = Q_{rev}/T = \Delta H/T = (-40.67 \text{ kJ mol-1})/(373.15 \text{ K}) = 109.0 \text{ J mol-1 K-1}$

$$\Delta G = \Delta H - T\Delta S = 0$$

$$\Delta A = \Delta G - P\Delta V = 3.10 \text{ kJ mol-1}$$

8.

$$C_{Pm} = a + bT - \frac{c'}{T^2}$$

$$\Delta H_m = \Delta H_{m0} + \int\left[\Delta a + (\Delta b)T - \frac{\Delta c'}{T^2}\right]dT$$

$$= \Delta H_{m0} + (\Delta a)T + \frac{\Delta b}{2}T^2 + \frac{\Delta c'}{T}$$

$$-\frac{\Delta H_m}{T^2} = -\frac{\Delta H_{m0}}{T^2} - \frac{\Delta a}{T} - \frac{\Delta b}{2} + \frac{\Delta c'}{T^3}$$

$$\frac{\Delta G_m}{T} = I + \int\left[-\frac{\Delta H_{m0}}{T^2} - \frac{\Delta a}{T} - \frac{\Delta b}{2} + \frac{\Delta c'}{T^3}\right]dT$$

$$= I + \frac{\Delta H_{m0}}{T} - \Delta a \ln T - \frac{\Delta b}{2}T - \frac{\Delta c'}{2T^2}$$

$$\Delta Y_m = -\frac{\Delta G_m}{T}$$

$$= -I - \frac{\Delta H_{m0}}{T} + (\Delta a) \ln T + \frac{\Delta b}{2}T + \frac{\Delta c'}{2T^2}$$

10.   a) Equation 7-121 can be expressed as

$$G_e = KT^{-1/2}$$

$$S_e = -\left(\frac{\partial G_e}{\partial T}\right)_P = 1/2\ KT^{-3/2} = \frac{1}{2T}KT^{-1/2} = \frac{G_e}{2T}$$

$$H_e = G + TS = KT^{-1/2} + T\frac{1}{2T}KT^{-1/2} = 3/2\,G_e$$

b) The negative value of S_e means that the number of arrangements of charged particles in a solution is less than the number of arrangements of uncharged particles. This difference in the number of arrangements results from the strong attraction of the polar solvent molecules to the charged particles.

12.   a) If we consider G as a function of P and n,

$$dG = \left(\frac{\partial G}{\partial n}\right)_P dn + \left(\frac{\partial G}{\partial P}\right)_n dP$$

$$\left(\frac{\partial G}{\partial n}\right)_V = \left(\frac{\partial G}{\partial n}\right)_P + \left(\frac{\partial G}{\partial P}\right)_n \left(\frac{\partial P}{\partial n}\right)_V$$

$$= \left(\frac{\partial G}{\partial n}\right)_P + V\left(\frac{\partial P}{\partial n}\right)_V$$

b)

$$G = A + PV$$

$$\left(\frac{\partial G}{\partial n}\right)_V = \left(\frac{\partial A}{\partial n}\right)_V + V\left(\frac{\partial P}{\partial n}\right)_V$$

$$\left(\frac{\partial G}{\partial n}\right)_P + V\left(\frac{\partial P}{\partial n}\right)_V = \left(\frac{\partial A}{\partial n}\right)_V + V\left(\frac{\partial P}{\partial n}\right)_V$$

$$\left(\frac{\partial A}{\partial n}\right)_V = \left(\frac{\partial G}{\partial n}\right)_P$$

14.   a) $Q_{rev} = T\Delta S$, so that we need to calculate ΔS.

$$\Delta S = -\,[\partial(\Delta G)/\partial T]_P = n\mathcal{F}(\partial\mathcal{E}/\partial T)_P$$

$$= (1)(9.648\times10^4 \text{ C mol}^{-1})(-1\times10^{-4} \text{ V K}^{-1})$$

$$= -9.648 \text{ J mol}^{-1} \text{ K}^{-1}$$

$$Q_{rev} = T\Delta S = (298.15 \text{ K})(-9.648 \text{ J mol}^{-1} \text{ K}^{-1})$$

$$= -2.88 \text{ kJ mol}^{-1}$$

Thus, heat is transferred from the cell to the thermostat.

b) In order to calculate Q for the irreversible case, we need to obtain ΔU and therefore W for the

reversible case. (We assume no PV work is done.)

$$W_{rev} = \Delta G = -n\mathscr{F}\mathscr{E} = -(1)(9.648 \times 10^4 \text{ C mol}^{-1})(-0.100 \text{ V})$$

$$= 9.648 \text{ kJ mol}^{-1} \ (1 \text{ VC} = 1 \text{ J})$$

$$\Delta U = Q_{rev} + W_{rev} = -2.88 \text{ kJ mol}^{-1} + 9.648 \text{ kJ mol}^{-1}$$

$$= 6.77 \text{ kJ mol}^{-1}$$

For the reverse process, with no work done,

$$Q_{irrev} = \Delta U = -6.77 \text{ kJ mol}^{-1}$$

Thus, heat is also transferred from the cell to the thermostat when the cell is short circuited and

no work is done. Remember that the second law states that $\Delta S \geq Q/T$, and ΔS for the reverse

process is 9.648 J mol^{-1} K^{-1}, which is greater algebraically than Q/T.

16.

$$\Delta G_m (T,P) = \Delta G_m (T_0,P_0) + \int_{T_0,P_0}^{T,P_0} (-\Delta S_m) dT + \int_{T,P_0}^{T,P} (\Delta V_m) dP$$

$$= \Delta G_m (T_0,P_0) + \int_{T_0,P_0}^{T,P_0} \left[-\Delta S_m(T_0,P_0) - \int_{T_0,P_0}^{T,P_0} \frac{\Delta C_{Pm}}{T} dT \right] dT$$

$$+ \int_{T,P_0}^{T,P} \left[\Delta V_m(T_0,P_0) + \int_{T_0,P_0}^{T,P_0} \left(\frac{\partial \Delta V_m}{\partial T} \right)_P dT + \int_{T,P_0}^{T,P} \left(\frac{\partial \Delta V_m}{\partial P} \right)_T dP \right] dP$$

$$= \Delta G_m(T_0,P_0) - \Delta S_m(T_0,P_0)(T - T_0) - \Delta C_{Pm}\left[T\ln\frac{T}{T_0} + T_0 - T\right]$$

$$+ \Delta V_m(T_0,P_0)(P - P_0) + \left(\frac{\partial\Delta V_m}{\partial T}\right)_P (T - T_0)(P - P_0) + \left(\frac{\partial\Delta V_m}{\partial P}\right)_T\frac{(P - P_0)^2}{2}$$

$$= 10600 \text{ J mol}^{-1} - (950 \text{ J mol}^{-1} \text{ K}^{-1})(35\text{K})$$

$$- (16000 \text{ J mol}^{-1} \text{ K}^{-1})\left[308.15 \text{ K ln} \frac{308.15\text{K}}{273.15\text{K}} - 35\text{K}\right]$$

$$+ (- 14.3\times10^{-6} \text{ m}^3 \text{ mol}^{-1})(300\times10^6 - 101.325\times10^3) \text{ J m}^{-3}$$

$$+ (1.32\times10^{-6} \text{ m}^3 \text{ mol}^{-1} \text{ K}^{-1})(35\text{K})((300\times10^6 - 101.325\times10^3) \text{ J m}^{-3}$$

$$+ (- 0.296\times^{-12} \text{ m}^6 \text{ J}^{-1} \text{ mol}^{-1})[(300\times10^6 - 101.325\times10^3) \text{ J m}^{-3}]^2/2$$

$$= 19110 \text{ J mol}^{-1}$$

18.   a)

$$\log_{10} K = 8.188 - \frac{2315.5 \text{ K}}{T} - 0.01025 \text{ K}^{-1} T$$

$$\ln K = 18.854 - \frac{5331.6 \text{ K}}{T} - 0.02360 \text{ K}^{-1} T$$

$$\Delta G_m^\circ = - RT \ln K$$

$$= - 18.854RT + (5331.6 \text{ K})R + (0.02360 \text{ K}^{-1}) RT^2$$

$$\Delta Y_m^\circ = - \frac{\Delta G_m^\circ}{T} = 18.854R - \frac{(5331.6 \text{ K})R}{T} - (0.02360 \text{ K}^{-1})RT$$

At 25°C,

$$\Delta G_m^\circ = (-\ 18.854)(8.3145 \text{ J mol}^{-1} \text{ K}^{-1})(298.15 \text{ K})$$

$$+ \ (5331.6 \text{ K})(8.3145 \text{ J mol}^{-1} \text{ K}^{-1})$$

$$+ \ (0.02360 \text{K}^{-1})(8.3145 \text{ J mol}^{-1} \text{ K}^{-1})(298.15 \text{ K})^2 = 15.0 \text{ kJ mol}^{-1}$$

$$\Delta Y_m^\circ = -\ \frac{\Delta G_n^\circ}{T} = -\ \frac{15.0 \text{ kJ mol}^{-1}}{298.15 \text{K}} = -\ 5.03 \text{ kJ mol}^{-1} \text{ K}^{-1}$$

b)

$$\Delta Y_m^\circ = R\ln K \qquad \text{and}$$

$$\left(\frac{\partial \ln K}{\partial T}\right)_P = \frac{\Delta H_m^\circ}{RT^2} \qquad \text{so that}$$

$$\Delta H_m^\circ = T^2\left(\frac{\partial \Delta Y_m^\circ}{\partial T}\right)_P$$

$$\Delta H_m^\circ = (5331.6 \text{ K})R - (0.02360 \text{ K}^{-1})RT^2$$

c) At 25°C

$$\Delta H_m^\circ = [5331.6 \text{ K} - (0.02360 \text{ K}^{-1})(298.15 \text{ K})^2](8.3145 \text{ J mol}^{-1} \text{ K}^{-1})$$

$$= 26.9 \text{ kJ mol}^{-1}$$

d)

$$\Delta S_m^\circ = \frac{\Delta H_m^\circ - \Delta G_m^\circ}{T} = \frac{(26.9 - 15.0) \text{ kJ mol}^{-1}}{298.15 \text{ K}}$$

$$= 39.7 \text{ J mol}^{-1} \text{ K}^{-1}$$

e) $\Delta C_{Pm} = (\partial \Delta H_m^\circ / \partial T)_P = -\ 2(0.02360 \text{ K}^{-1})RT$

$$= -\ 2(0.02360 \text{ K}^{-1})(8.3145 \text{ J mol}^{-1} \text{ K}^{-1})(298.15 \text{ K}) = 117 \text{ J mol}^{-1} \text{ K}^{-1}$$

Since ΔH is a linear function of T, ΔC_{Pm} is a constant.

22.   By definition,

$$\mu_{JT} = \left(\frac{\partial T}{\partial P}\right)_H = - \frac{\left(\frac{\partial H}{\partial P}\right)_T}{\left(\frac{\partial H}{\partial T}\right)_P} = - \frac{\left(\frac{\partial H}{\partial P}\right)_T}{C_P}$$

In order to find an expression for $(\partial H/\partial P)_T$, we start with the total differential of Y as a function

of T and P.

$$dY = \frac{H}{T^2}dT - \frac{V}{T}dP$$

Since Y is a state function,

$$\frac{1}{T^2}\left(\frac{\partial H}{\partial P}\right)_T = - \frac{1}{T}\left(\frac{\partial V}{\partial T}\right)_P + \frac{V}{T^2}$$

and,

$$\left(\frac{\partial H}{\partial P}\right)_T = - T\left(\frac{\partial V}{\partial T}\right)_P + V$$

and,

$$\mu_{JT} = \frac{1}{C_P}\left[T\left(\frac{\partial V}{\partial T}\right)_P - V\right]$$

Chapter 8

Application of the Gibbs Function and the Planck Function to Some Phase Changes

2.   a) The desired transformation is

$$H_2O(\ell,100°C,1\ Bar) = H_2O(g,100°C,1\ Bar)$$

This transformation can be divided into three steps for ease of calculation.

1) $H_2O(\ell,100°C,1\ Bar) = H_2O(\ell,100°C,101.325\ kPa)$

2) $H_2O(\ell,100°C,101.325\ kPa) = H_2O(g,100°C,101.325\ kPa)$

3) $H_2O(g,100°C,101.325\ kPa) = H_2O(g,100°C,100.000\ kPa)$

$$\Delta G_{m1} = \int_{100kPa}^{101.325kPa} V_m dP = V_m(101.325 - 100.000)\ kPa$$

$$= (18.797\times10^{-6}\ m^3\ mol^{-1})(1.325\times10^3\ J\ m^{-3}) = 0.0249\ J\ mol^{-1}$$

$$\Delta G_{m2} = 0(equilibrium)$$

$$\Delta G_{m3} = RT \ln (P_2/P_1)$$

$$= (8.3145\ J\ mol^{-1}\ K^{-1})(373.15K)\ \ln \frac{100.00\ kPa}{101.325\ kPa} = -40.839\ J\ mol^{-1}$$

$$\Delta G_m^° = \Delta G_{m1} + \Delta G_{m2} + \Delta G_{m3} = -40.814\ J\ mol^{-1}$$

$$\Delta Y_m^° = -\frac{\Delta G_m^°}{T} = \frac{-40.814\ J\ mol^{-1}}{373.15\ K} = 0.10938\ J\ mol^{-1}\ K^{-1}$$

b) The desired transformation is

$$H_2O(\ell,25°C,1\ Bar) = H_2O(g,25°C,1\ Bar)$$

This transformation can also be divided into three steps for ease of calculation.

$$1)\ H_2O(\ell,25°C,1\ Bar) = H_2O(\ell,25°C,3.17\ kPa)$$

$$2)\ H_2O(\ell,25°C,3.17\ kPa) = H_2O(g,25°C,3.17\ kPa)$$

$$3)\ H_2O(g,25°C,3.17\ kPa) = H_2O(g,25°C,100.000\ kPa)$$

$$\Delta G_{m1} = V_m(P_2 - P_1)$$

$$= (18.068 \times 10^{-6}\ m^3\ mol^{-1})(3.17 \times 10^3 - 100.000 \times 10^3)\ J\ m^{-3} = -1.750\ J\ mol^{-1}$$

$$\Delta G_{m2} = 0 (equilibrium)$$

$$\Delta G_{m3} = RT\ \ln\ (P_2/P_1)$$

$$= (8.3145\ J\ mol^{-1}\ K^{-1})(298.15\ K)\ \ln\ \frac{100.00 kPa}{3.17 kPa} = 8556\ J\ mol^{-}$$

$$\Delta G_m^{\circ} = \Delta G_{m1} + \Delta G_{m2} + \Delta G_{m3} = 8554\ J\ mol^{-1}$$

$$\Delta Y_m^{\circ} = \frac{\Delta G_m^{\circ}}{T} = \frac{8554\ J\ mol^{-1}}{298.15\ K} = 0.28690\ J\ mol^{-1}\ K^{-1}$$

4.   a)

$$\left(\frac{\partial \Delta S_m}{\partial T}\right)_P = \frac{\Delta C_{Pm}}{T}$$

$$\Delta S_m = \Delta S_{m0} + \int \frac{\Delta C_{Pm}}{T} dT$$

$$= \Delta S_{m0} + \Delta C_{Pm}\ \ln\ (T/K)$$

where ΔS_{m0} is a constant of integration, and ΔC_{Pm} is assumed independent of temperature.

$$\left(\frac{\partial \Delta G_m}{\partial T}\right)_P = - \Delta S_m = - \Delta S_{m0} - \Delta C_{Pm} \ln (T/K)$$

$$\Delta G_m = I' + \int [- \Delta S_{m0} - \Delta C_{Pm} \ln (T/K)] dT$$

$$= I' + (\Delta C_{Pm} - \Delta S_{m0})T - \Delta C_{pm}[T \ln (T/K)]$$

where I' is a constant of integration.

b) For $H_2O(\ell, 0°C) = H_2O(s, 0°C)$

$\Delta H_m = - 6008$ J mol^{-1}, and

$\Delta S_m = \Delta H_m /T = (-6008$ J mol$^{-1})/(273.15$ K$) = -21.995$ J mol^{-1} K^{-1}

$\Delta C_{Pm} = 38.9$ J mol^{-1} K^{-1}

Thus, $\Delta S_{m0} = \Delta S_m - \Delta C_{Pm} \ln T$

$$= [- 21.995 - 38.9 \ln (273.15)]$$ J mol^{-1} K^{-1} $= - 240.225$ J mol^{-1} K^{-1}

At 0°C, $\Delta G_m = 0$

Thus, $I' = \Delta G_m - (\Delta C_{Pm} - \Delta S_{m0})T - \Delta C_{Pm} (T \ln T)$

$$= 0 - (38.9 + 240.225)$$J mol^{-1} K^{-1} $(273.15$ K$)$

$$+ (38.9$$ J mol^{-1} K$^{-1})(273.15$ K$)(\ln 273.15) = -16634$ J mol^{-1}

c) At - 10°C,

$$\Delta G_m = I' + (\Delta C_{Pm} - \Delta S_{m0})T - \Delta C_{Pm}(T \ln T) = - 228$$ J mol^{-1}

compared with - 213 J mol^{-1} by the other methods.

6.   For the reaction

Graphite = Diamond

$\Delta S_m = (2.377 - 5.740)$J mol^{-1} K^{-1}

$\Delta V_m = [(12.00$ g mol$^{-1})/(3.51$ g cm$^{-3}) - (12.00$ g mol$^{-1})/(2.22$ g cm$^{-3})](10^{-6}$ m^3 cm$^{-3})$

$$= -2.00 \times 10^{-6} \text{ m}^3 \text{ mol}^{-1}$$

$$\Delta G_m = \Delta G_m (298.15 \text{ K}, 1 \text{ Bar}) + \int_{298.15 \text{ K}}^{T} (-\Delta S_m) dT + \int_{1 \text{ Bar}}^{P} (\Delta V_m) dP$$

$$= [2900 \text{ J mol}^{-1} + 3.363 \text{ J mol}^{-1} \text{ K}^{-1}(T - 298.15 \text{ K})] - 2.00 \times 10^{-6} \text{ m}^3 \text{ mol}^{-1}(P - 1 \text{ Bar})$$

We conclude from the value of ΔS_m that ΔG_m would become more negative at lower temperatures, and we conclude from the value of ΔV_m that ΔG_m would become more negative at higher pressures. Of course, at lower temperatures, conversion of graphite to diamond would be infinitely slow even if the calculated ΔG_m were negative. Thus, let us consider a temperature of 3000 K (see Figure 13-2), at which the transition **might** occur at a reasonable rate, and calculate the pressure at which ΔG_m would equal 0, the minimum pressure for a negative value of ΔG_m. (1 Bar will be negligible with respect to this pressure.)

$$0 = [2900 \text{ J mol}^{-1} + (3.363 \text{ J mol}^{-1} \text{ K}^{-1})(3000 \text{ K} - 298.15 \text{ K})] - (2.00 \times 10^{-6} \text{ m}^3 \text{ mol}^{-1}) P$$

$$P = [2900 \text{ J mol}^{-1} \text{ K}^{-1} + (3.363 \text{ J mol}^{-1} \text{ K}^{-1})(3000 \text{ K} - 298.15 \text{ K})]/(2.00 \times 10^{-6} \text{ m}^3 \text{ mol}^{-1})$$

$$= 5.99 \times 10^9 \text{ J m}^{-3} = 5.99 \times 10^9 \text{ Pa} = 5.99 \times 10^4 \text{ Bar}$$

8.

$$\log_{10} P = 7.01667 - \frac{1321.331 \text{ K}}{t + 224.513 \text{ K}}$$

$$= 7.0166 - \frac{1321.331 \text{ K}}{T - 48.637 \text{ K}}$$

$$\ln P = 16.1565 - \frac{3042.477 \text{ K}}{T - 48.637 \text{ K}}$$

$$\frac{d \ln P}{dT} = \frac{3042.477 \text{ K}}{(T - 48.637 \text{ K})^2} = \frac{\Delta H_m}{RT^2}$$

$$\Delta H_m = \frac{(3042.477 \text{ K}) RT^2}{(T - 48.637 \text{ K})^2}$$

$$= \frac{(3042.477 \ K)(8.3145 \ J \ mol^{-1} \ K^{-1})(298.15 \ K)^2}{(298.15 \ K \ - \ 48.637 \ K)^2}$$

$$= 36.119 \ kJ \ mol^{-1}$$

10.   The desired transformation is

$$H_2O(\ell,- 5°C) = H_2O(s,- 5°C)$$

This transformation can be expressed as the following series of steps for ease of calculation, since ΔG_m is a state function.

1) $H_2O(\ell,- 5°C, 1 \ Bar) = H_2O(\ell,- 5°C, 421.7 \ Pa)$

2) $H_2O(\ell,- 5°C, 421.7 \ Pa) = H_2O(g,- 5°C, 421.7 \ Pa)$

3) $H_2O(g,- 5°C, 421.7 \ Pa) = H_2O(g,- 5°C, 401.6 \ Pa)$

4) $H_2O(g,- 5°C, 401.6 \ Pa) = H_2O(s,- 5°C, 401.6 \ Pa)$

5) $H_2O(s,- 5°C, 401.6 \ Pa) = H_2O(s,- 5°C, 1 \ Bar)$

$$\Delta G_{m1} = V_m(\ell)(P_2 - P_1)$$

$$= (18.023 \times 10^6 \ m^3 \ mol^{-1})(421.7 - 100 \times 10^3) J \ m^{-3} = - 1.7947 \ J \ mol^{-1}$$

$$\Delta G_{m2} = 0 \ (equilibrium)$$

$$\Delta G_{m3} = RT \ ln \ (P_2/P_1)$$

$$= (8.3145 \ J \ mol^{-1} \ K^{-1})(268.15 \ K) \ ln \ [(401.6 \ Pa)/(421.7 \ Pa)] = - 108.9 \ J \ mol^{-1}$$

$$\Delta G_{m4} = 0 \ (equilibrium)$$

$$\Delta G_{m5} = V_m(s)(P_2 - P_1)$$

$$= (19.65 \times 10^{-6} \ m^3 \ mol^{-1})(100 \times 10^3 - 401.6) \ J \ m^{-3} = 1.957 \ J \ mol^{-1}$$

$$\Delta G_m° = \Delta G_{m1} + \Delta G_{m2} + \Delta G_{m3} + \Delta G_{m4} + \Delta G_{m5} = - 105.1 \ J \ mol^{-1}$$

$$\Delta Y_m° = - (\Delta G_m°/T) = - (- 105.1 \ J \ mol^{-1})/(268.15 \ K) = 0.3919 \ J \ mol^{-1} \ K^{-1}$$

12.   Rhombic sulfur must have a lower value of G_m than monoclinic sulfur or liquid sulfur below

99.5°C because it is the stable phase below 99.5°C. Monoclinic sulfur must have a lower value

of G_m than rhombic sulfur or liquid sulfur in a range above 99.5°C in which it is the stable form.

Above 120°C the liquid must have the lowest value of G_m. In light of the diagram below, based

on the given data, a report that rhombic sulfur melts at 77°C should be viewed skeptically.

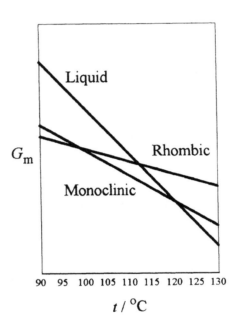

Figure 8-1. Sketch of G_m as a function of t for rhombic, monoclinic, and liquid sulfur.

14.   If we fit the vapor pressure data for mercury to a linear least squares expression for ln P against

1/T, the result is

$$\ln P = 16.22 \pm .02 - [(7281 \pm 7 \text{ K})/T]$$

From this expression,

$$-7281 \text{ K} = (\Delta H_m)/zR)$$

$$(\Delta H_m)/z = (7281 \text{ K})(8.3145 \text{ J mol}^{-1} \text{ K}^{-1})$$

$$= 60.53 \text{ kJ mol}^{-1}$$

Since the data were not obtained at evenly spaced values of the independent variable, we cannot use the simple form of the Savitsky and Golay algorithm for numerical differentiation described in Chapter 23 in order to determine whether the derivative of the experimental data in this form is constant. As an alternative we can fit the data nonlinearly to the Antoine equation, and the result is

$$\ln P = 15.52 \pm 0.11 - [6667 \pm 97 \text{ K } (1/T)]/[1 - (19 \pm 3 \text{ K})(1/T)]$$

The graphs for the two fits cannot be distinguished visually on the scale of the illustrations below.

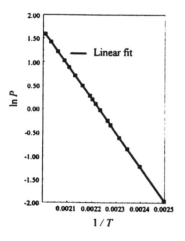

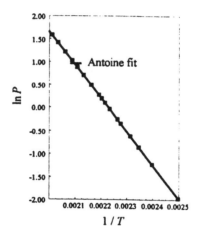

Figure 8-2. Linear fit of ln P against $1/T$.   **Figure 8-3.** Antoine fit of ln P against $1/T$.

One way to decide which equation is the better representation is to calculate the residuals, the differences between experimental and calculated values for the two equations and plot them against the independent variable. The results are shown below.

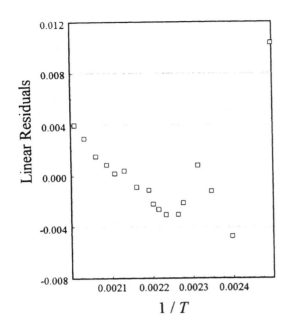

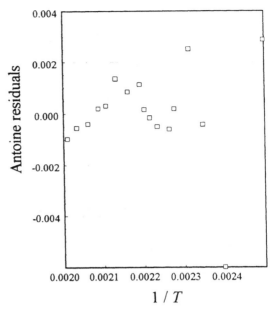

Figure 8-4. Residuals for the linear fit.   **Figure 8-5.** Residuals for the Antoine fit.

The residuals for the linear fit are seen to be nonrandomly dependent on the independent variable,

whereas the residuals for the nonlinear fit are relatively random with respect to the value of the

independent variable and smaller. We can then conclude that the Antoine equation is a better fit

of the data than the linear fit.

If we use the Antoine equation to calculate $\Delta H_m/z$, the result is

$$\frac{d\ln P}{d(1/T)} = -\frac{\Delta H_m}{zR} = -T^2 \frac{6667\text{K}}{(T - 19\text{ K})^2}$$

At 450 K, the center of the range of temperatures,

$$\frac{\Delta H_m}{z} = \frac{(8.3145 \text{ J mol}^{-1}\text{ K}^{-1})(450\text{ K})^2(6663\text{K})}{(431\text{ K})^2}$$

$$= 60.4 \text{ kJ mol}^{-1}$$

in good agreement with the value calculated from the linear fit. Although $\Delta H_m/z$ can be calculated at each experimental point, ΔH_m, and hence ΔC_{Pm}, cannot be calculated without information about z as a function of T and P. Since the linear fit is quite close, it is a good assumption that ΔC_{Pm} is close to zero.

16.   Table 8-1 contains the data obtained from the NIST WEBBOOK and the quantities calculated from those data in a spreadsheet in PSIPLOT6, the graphics program used to produce the graphs.

a)

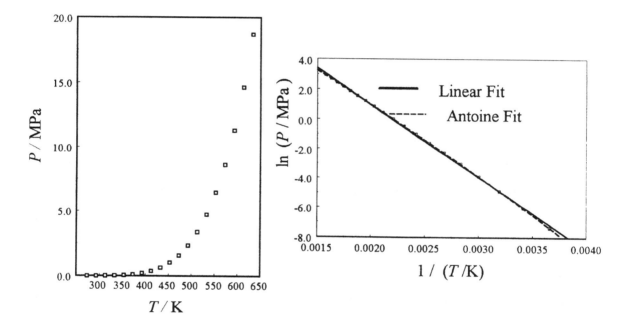

Figure 8-6. A plot of the vapor pressure of water   **Figure 8-7.** A plot of ln P against $1/T$ for water.

as a function of temperature

The curves in Figure 8-7 show clearly that the linear plot is not a good fit, whereas the Antoine equation is a good fit. In this case, it is not necessary to use residuals to distinguish which curve is a better fit.

b) The column in Table 8-1 labelled dP/dT shows the values calculated from the analytical derivative of the best-fit Antoine equation.

c) The column in Table 8-1 labelled $\Delta S/\Delta V$ shows the values of $(S_g - S_\ell)/(V_g - V_\ell)$ from the WEBBOOK. They compare favorably with the values of dP/dT.

d) The values of $\Delta H/Z$ in Table 8-1 were calculated from the values of dP/dT from the equation above Equation 8-17. The values of Z were obtained from the Redlich-Kwong equation, Equation 5-57, with the values of a and b calculated from Equation 5-58, with critical constants from Table 5-3. Values of V_{mg} were from the WEBBOOK. The resultant values of ΔH calculated are in good agreement with the values from the WEBBOOK.

Data for Table 8-1 taken with permission from:

E. W. Lemmon, M. O. McLinden and D. G. Friend, Thermophysical Properties of Fluid Systems" in **NIST Chemistry WebBook, NIST Standard Reference Database Number 69**, Eds. W. G. Mallard and P. J. Linstrom, November 1998, National Institute of Standards and Technology, Gaithersburg, MD 20899 (http://webbook.nist.gov).

This data is also available in:

A. Pruss and W. Wagner, **J. Phys. Chem. Reference Data**, 1998.

Table 18-1. Spreadsheet for Exercise 8-16.

T/K	P/MPa	$V_{m\ell}/\ell$ mol^{-1}	V_{mg}/ℓ mol$_{-1}$	$H_{m\ell}/$kJ mol^{-1}	$H_{mg}/$kJ mol^{-1}
273.16	0.00061165	0.018019	3711.0	0.000	45.055
293.16	0.0023408	0.018048	1039.9	1.5125	45.713
313.16	0.0073889	0.018157	351.39	3.0189	46.363
333.16	0.019956	0.018324	138.07	4.5258	46.999
353.16	0.047434	0.018539	61.322	6.0361	47.615
373.16	0.10145	0.018798	30.107	7.5522	48.201
393.16	0.19874	0.019102	16.051	9.0771	48.748
413.16	0.36164	0.019452	9.1575	10.615	49.244
433.16	0.61839	0.019853	5.5254	12.17	49.676
453.16	1.00300	0.020311	3.4914	13.747	50.032
473.16	1.5553	0.020835	2.2913	15.335	50.299
493.16	2.3200	0.021441	1.5507	17.000	50.460
513.16	3.3475	0.022149	1.0754	18.693	50.496
533.16	4.693	0.02299	0.75963	20.447	50.381
553.16	6.4176	0.024012	0.54312	22.284	50.080
573.16	8.5891	0.025298	0.39015	24.232	49.535
593.16	11.2860	0.027007	0.27866	26.343	48.651
613.16	14.6030	0.029503	0.19418	28.727	47.232
633.16	18.6680	0.03415	0.12516	31.379	44.703

Table 18-1. (continued)

$S_{m\ell}/$	$S_{mg}/$	ln P/MPa	$1/(T/K)$	ln P	ln P
J mol^{-1} K^{-1}	J mol^{-1} K^{-1}			linear fit	Antoine fit
0.0000	164.94	-7.39935	0.0036609	-7.20327	-7.39779
5.3438	156.12	-6.05726	0.0034111	-5.97667	-6.05327
10.314	148.72	-4.90778	0.0031933	-4.90676	-4.90642
14.978	142.46	-3.91423	0.0030016	-3.96530	-3.91664
19.379	137.11	-3.04842	0.0028316	-3.13047	-3.05372
23.552	132.48	-2.28819	0.0026798	-2.38513	-2.29475
27.528	128.43	-1.61576	0.0025435	-1.71562	-1.62201
31.334	124.83	-1.01711	0.0024204	-1.11093	-1.02159
34.998	121.59	-0.48064	0.0023086	-0.56208	-0.48242
38.541	118.69	0.00300	0.0022067	-0.06167	0.00440
41.987	115.84	0.44167	0.0021135	0.39643	0.44616
45.358	113.21	0.84157	0.0020277	0.81737	0.84883
48.679	110.65	1.20821	0.0019487	1.20551	1.21738
51.975	108.12	1.54607	0.0018756	1.56452	1.55597
55.281	105.53	1.85904	0.0018078	1.89757	1.86811
58.644	102.79	2.15049	0.0017447	2.20738	2.15678
62.144	99.753	2.42356	0.0016859	2.49630	2.42454
65.940	96.120	2.68123	0.0016309	2.76637	2.67358
70.563	91.038	2.92681	0.0015794	3.01938	2.90579

Table 8-1. (continued)

$H_{vap}/$ mol^{-1}	$\Delta S_{vap}/$ J mol^{-1} K^{-1}	$\Delta V_{vap}/(\ell/\text{ mol})$	$\Delta S/\Delta V$	dP/dt
5055	164.94	3711.0	0.04445	0.05964
4200	150.78	1039.9	0.14499	0.14094
3344	138.41	351.37	0.39390	0.39527
2473	127.48	138.05	0.92344	0.92394
1578	117.73	61.303	1.92046	1.92156
0648	108.93	30.088	3.62029	3.62229
9670	100.90	16.032	6.29383	6.29637
8629	93.50	9.1380	10.23151	10.23487
7506	86.59	5.5055	15.72814	15.72865
6285	80.15	3.4711	23.09045	23.07443
4964	73.85	2.2705	32.52770	32.53249
3460	67.85	1.5293	44.36920	44.37080
1803	61.97	1.0533	58.83783	58.84189
9934	56.15	0.73664	76.21769	76.24217
7796	50.25	0.51911	96.79874	96.73833
5303	44.15	0.36485	120.99701	121.25876
2308	37.61	0.25165	149.44785	148.53397
8505	30.18	0.16468	183.26785	186.75432
3324	20.48	0.09101	224.97528	211.37474

Table 8-16. (continued)

$(d \ln P)/(d\ 1/T)$	$\Delta H_{vap}/Z$	Z	ΔH
-5396	45306	0.99446	45055
-5330	44231	0.99931	44201
-5227	43324	1.00046	43344
-5136	42549	0.99821	42473
-5049	41879	0.99283	41579
-4969	41294	0.98437	40649
-4895	40779	0.97282	39671
-4829	40322	0.95800	38629
-4771	39914	0.93966	37506
-4723	39548	0.91750	36285
-4682	39216	0.89157	34964
-4650	38916	0.85981	33460
-4628	38641	0.82303	31803
-4616	38390	0.77973	29934
-4615	38160	0.72842	27796
-4628	37947	0.66680	25303
-4656	37750	0.59094	22308
-4729	37567	0.49258	18505
-4785	37397	0.35628	13324

Chapter 9

The Third Law of Thermodynamics

2.

$$\Delta S = \Delta S_{0K} + \int_0^T \frac{\Delta C_V}{T} dT$$

Since ΔS must be finite at all temperatures, the integral on the right must converge. If $\Delta C_V \neq 0$ when $T = 0$, the integrand would become infinite at the limit of $T = 0$, and ΔS would not be finite. Thus

$$\lim_{T \to 0} \Delta C_V = 0$$

4. The Debye equation gives the following equation for the entropy of each form of methylammonium chloride at the lowest temperature at which heat capacity is obtained:

$$S_m(T) = \{C_P(T)\}/3$$

Thus for the β form,

$$S_m(12.0 \text{ K}) = (0.845 \text{ J mol}^{-1} \text{ K}^{-1})/3 = 0.282 \text{ J mol}^{-1} \text{ K}^{-1}$$

$$S_m(220.4 \text{ K}) - S_m(12.0 \text{ K}) = 93.412 \text{ J mol}^{-1} \text{ K}^{-1}$$

and

$$S_m(220.4 \text{ K}) = 93.694 \text{ J mol}^{-1} \text{ K}^{-1}$$

For the γ form,

$$S_m(19.5 \text{ K}) = (5.966 \text{ J mol}^{-1} \text{ K}^{-1})/3 = 1.989 \text{ J mol}^{-1} \text{ K}^{-1}$$

$$S_m(220.4 \text{ K}) - S_m(19.5 \text{ K}) = 99.918 \text{ J mol}^{-1} \text{ K}^{-1}$$

and

$$S_m(220.4 \text{ K}) = 101.907 \text{ J mol}^{-1} \text{ K}^{-1}$$

For the transition from β to γ,

$$\Delta S_m = (101.987 - 93.694) \text{ J mol}^{-1} \text{ K}^{-1} = 8.213 \text{ J mol}^{-1} \text{ K}^{-1}$$

Since $\Delta G_m = 0$ at the transition temperature,

$$\Delta H_m = T\Delta S_m = (220.4 \text{ K})(8.213 \text{ J mol}^{-1} \text{ K}^{-1}) = 1.810 \text{ kJ mol}^{-1}$$

6.   From heat capacity measurements for $Na_2SO_4 \cdot 10H_2O(s)$,

$$S_m^\circ(298.15 \text{ K}) = 585.55 \text{ J mol}^{-1} \text{ K}^{-1}$$

For the reaction

$$Na_2SO_4(s) + 10 \text{ H}_2O(g) = Na_2SO_4 \cdot 10H_2O(s)$$

$$\Delta G_m^\circ = -91.190 \text{ kJ mol}^{-1}$$

$$\Delta H_m^\circ = -521.950 \text{ kJ mol}^{-1}, \text{ and}$$

$$\Delta S_m^\circ = (\Delta H_m^\circ - \Delta G_m^\circ)/T = -1.4448 \text{ kJ mol}^{-1} \text{ K}^{-1}$$

From the standard entropies,

$$\Delta S_m^\circ = S_m^\circ(Na_2SO_4 \cdot 10H_2O) - S_m^\circ(Na_2SO_4) - 10S_m^\circ(H_2O)$$

and $S_m^\circ(Na_2SO_4 \cdot 10H_2O) = \Delta S_m^\circ + S_m^\circ(Na_2SO_4) + 10 \, S_m^\circ(H_2O)$

$$= -1444.8 \text{ J mol}^{-1} + 149.49 \text{ J mol}^{-1}$$

$$+ 10(188.715 \text{ J mol}^{-1})$$

$$= 591.9 \text{ J mol}^{-1}$$

The value of ΔS_m° of reaction does not depend on any assumption about S at 0 K. If the other standard entropies are correct, the value of $S_m°$ for $Na_2SO_4 \cdot 10H_2O$ calculated from heat capacity measurements on the assumption that S is zero at 0 K is too small by 6.3 J mol^{-1} K^{-1}. Thus $S_m°$ is 6 J mol^{-1} K^{-1} at 0 K, and it is not a "perfect crystalline substance" at 0 K.

8.

$$\lim_{T \to 0} \Delta Y_m = \lim_{T \to 0} - \frac{\Delta G_m}{T}$$

Since ΔG_m and T are both equal to zero at 0 K, the ratio on the right is an indeterminate form. Therefore, we can use L'Hôpital's theorem to evaluate the limit.

$$\lim_{T \to 0} - \frac{\Delta G_m}{T} = \lim_{T \to 0} \frac{\Delta S_m}{1} = 0$$

Chapter 10

Application of the Gibbs and the Planck Function to Chemical Changes

2.

$$\Delta G_m^\circ = \Delta H_m^\circ - T\Delta S_m^\circ = (-1200 + 1.2 \text{ K}^{-1} T) \text{ cal mol}^{-1}$$

$$\ln K = -\frac{\Delta G_m^\circ}{RT} = \frac{[(1200/T) - 1.2 \text{ K}^{-1}] \text{ cal mol}^{-1}}{1.987 \text{ cal mol}^{-1} \text{ K}^{-1}}$$

$$= \frac{603.9 \text{ K}}{T} - 0.603$$

4. a) **Table 10-1.**

T/K	492.6	522.0	555.0	580.0	613.0
% trans	2.69	3.61	5.09	6.42	8.23
K_{eq}	0.0276	0.0375	0.0536	0.0686	0.0897
$\ln K_{eq}$	-3.590	-3.283	-2.926	-2.679	-2.411
10^3 K/T	2.030	1.916	1.802	1.724	1.631
residuals	0.0144	-0.0209	-0.0044	0.0101	0.0007

b)

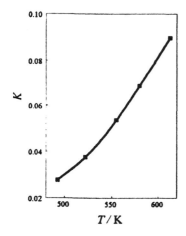

Figure 10-1. A plot of K against T.

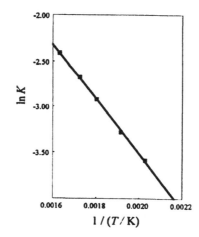

Figure 10-2. A plot of $\ln K$ against $1/T$.

The plot of ln K against $1/T$ appears to be linear. The best-fit linear least squares equation for the points is

$$\ln K = 2.466 - (2990\ K)\ \frac{1}{T}$$

The residuals above suggest that the dependence is not exactly linear, but we shall use the assumption of linearity to determine ΔH.

c) From the graph of ln K_{eq} against $1/T$ above, the slope is -2.990×10^3 K^{-1}.

$$\left(\frac{\partial \ln K}{\partial (1/T)} \right)_P = -T^2 \left(\frac{\partial \ln K}{\partial T} \right)_P = -T^2 \frac{\Delta H_m^\circ}{RT^2} = -\frac{\Delta H_m^\circ}{R}$$

Thus,

$$\Delta H_m^\circ = -R(\text{slope}) = -(8.3145\ \text{J mol}^{-1}\ \text{K}^{-1})(-2.990 \times 10^3\ \text{K}^{-1}) = 24.86\ \text{kJ mol}^{-1}$$

assumed constant over the temperature range.

At 555 K, $\Delta G_m^\circ = -RT \ln K_{eq} = -(8.3145\ \text{J mol}^{-1}\ \text{K}^{-1})(555\ \text{K})(-2.926) = 13.502\ \text{kJ mol}^{-1}$

$$\Delta S_m^\circ = (\Delta H_m^\circ - \Delta G_m^\circ)/T = [(24.86 - 13.502)\text{kJ mol}^{-1}]/(555\ \text{K}) = 20.46\ \text{J mol}^{-1}\ \text{K}^{-1}$$

assumed constant over the temperature range.

6. $Y = Y_A + Y_B$

$$= n_A Y_{mA}^\circ - n_A R\ \ln \frac{P_A}{P^\circ} + n_B Y_{mB}^\circ - n_B R\ \ln \frac{P_B}{P^\circ}$$

$$= n(1 - X)Y_{mA}^\circ - n(1 - X)R\ \ln \frac{(1 - X)P}{P^\circ} + n X Y_{mB}^\circ - n X R\ \ln \frac{X P}{P^\circ}$$

$$\left(\frac{\partial Y}{\partial X}\right)_{T,P} = n\,\Delta Y_m^\circ - n\,R\,\ln\frac{X}{1-X}$$

$$dY = n\left(\Delta Y_m^\circ - R\,\ln\frac{X}{1-X}\right)dX$$

a)

$$\Delta Y = \int_{X=0}^{X'} dY = \int_0^{X'}\left(n\,\Delta Y_m^\circ - nR\,\ln\frac{X}{1-X}\right)dX$$

$$= n\,X'\,\Delta Y_m^\circ - nR\int_0^{X'}\ln\frac{X}{1-X}\,dX$$

If we use integration by parts, with $u = \ln[X/(1-X)]$ and $dv = dX$,

$$\Delta Y = n\,X'\,\Delta Y_m^\circ - nR\,X\,\ln\frac{X}{1-X}\Big|_0^{X'} + nR\int_0^{X'} X\,d\,\ln\frac{X}{1-X}$$

By L'Hôpital's theorem, the limit of $\ln[X/(1-X)]$ as $X\to 0$ is 0, so that

$$\Delta Y = n\,X'\,\Delta Y_m^\circ - nR\,X'\,\ln\frac{X'}{1-X'} + nR\int_0^{X'}\frac{dX}{1-X}$$

$$= n\,X'\,\Delta Y_m^\circ - nR\left[X'\,\ln\frac{X'}{1-X'} + \ln(1-X')\right]$$

b) At equilibrium,

$$K = \frac{X'}{1-X'}, \quad X' = \frac{K}{1+K}, \quad 1-X' = \frac{1}{1+K}$$

$$\Delta Y = n\frac{K}{1+K}(R\,\ln K) - nR\frac{K}{1+K}\ln K - nR\,\ln\frac{1}{1+K}$$

$$= nR\,\ln(K+1)$$

8.   For reaction 1,

$$\Delta G_m^\circ = -1.44 \text{ kcal mol}^{-1} \text{ at } 173 \text{ K}$$

$$\ln K_{eq} = -\Delta G_m^\circ/(RT) = (1.44 \text{ kcal mol}^{-1})/[(1.987 \text{ kcal mol}^{-1})(173 \text{ K})] = 4.18$$

$$K_{eq} = 66, \text{ which is the desired ratio.}$$

For reaction 3,

$$\Delta G_m° = -0.89 \text{ kcal mol}^{-1} \text{ at } 173 \text{ K}$$

$$\ln K_{eq} = -\Delta G_m°/(RT) = (0.89 \text{ kcal mol}^{-1})/[(1.987 \text{ kcal mol}^{-1})(173 \text{ K}) = 2.6$$

$K_{eq} = 14$, which is the desired ratio.

For reaction 4,

$$\Delta G_m° = -1.62 \text{ kcal mol}^{-1} \text{ at } 173 \text{ K}$$

$$\ln K_{eq} = -\Delta G_m°/(RT) = (1.62 \text{ kcal mol}^{-1})/[(1.987 \text{ kcal mol}^{-1})(173 \text{ K}) = 4.45$$

$K_{eq} = 86$, which is the desired ratio.

10. $$\ln K_{eq} = 46.650 + 1.5 \ln T + (16.21 \times 10^9 \text{ K})/T$$

a) At 10^8 K, $\ln K_{eq} = 236.4$, $\log_{10} K_{eq} = 102.7$, $K_{eq} = 10^{103}$

b) At 10^8 K, $\Delta Y_m° = R \ln K_{eq} = -1.96 \text{ kJ mol}^{-1} \text{ K}^{-1}$

$$\Delta H_m° = RT^2 (d \ln K_{eq}/dT) = -2.01 \times 10^8 \text{ kJ mol}^{-1}$$

$$\Delta S_m° = \Delta H_m°/T + \Delta Y_m° = -4 \text{ J mol}^{-1} \text{ K}^{-1}$$

12.   For the conversion of Sn(gray) to Sn(white)

a) $$\Delta H_m° = \Delta_f H_m°(\text{white}) - \Delta_f H_m°(\text{gray}) = [0 - (-2090)] \text{J mol}^{-1} = 2090 \text{ J mol}^{-1}$$

$$\Delta S_m° = S_m°(\text{white}) - S_m°(\text{gray}) = (51.55 - 44.14) \text{J mol}^{-1} \text{ K}^{-1} = 7.41 \text{ J mol}^{-1} \text{ K}^{-1}$$

$$\Delta G_m° = \Delta H_m° - T\Delta S_m° = -119 \text{ J mol}^{-1}$$

$$\mathscr{E}° = -(\Delta G_m°)/(n\mathscr{F}) = (119 \text{ J mol}^{-1})/[2(9.6485 \times 10^4 \text{ C mol}^{-1})]$$

$$= 6.19 \times 10^{-3} \text{ J C}^{-1} = 6.19 \times 10^{-3} \text{ V C C}^{-1} = 6.19 \times 10^{-3} \text{ V}$$

b) $$W_{net,rev} = \Delta G_m° = -119 \text{ J mol}^{-1}$$

$$\Delta H_m = \Delta H_m° = 2090 \text{ J mol}^{-1}$$

$$Q_{rev} = T\Delta S_m = T\Delta S_m° = (298.15 \text{ K})(7.41 \text{ J mol}^{-1}) = 2209 \text{ J mol}^{-1} \text{ K}^{-1}$$

$$\Delta U_m = \Delta H_m - P\Delta V_m$$

$$\Delta V_m = \{[118.69)/(7.31)] - [(118.69)/(5.75)]\}[10^{-6} \text{ m}^3 \text{ cm}^{-3} \text{ (g mol-1)/(g cm}^{-3})]$$

$$= -0.440 \times 10^{-5} \text{ m}^3 \text{ mol-1}$$

$$P\Delta V_m = 0.45 \text{ J mol}^{-1}$$

$$\Delta U_m \approx \Delta H_m = 2090 \text{ J mol}^{-1}$$

c) $$W_{net} = 0 \; ; \; W \approx W_{net} = 0$$

$$Q = \Delta U_m - W = \Delta U_m = 2090 \text{ J mol}^{-1} = \Delta H_m$$

$$\Delta S_m = 7.41 \text{ J mol}^{-1} \text{ K}^{-1}$$

$$\Delta G_m = -119 \text{ J mol}^{-1}$$

14.   For the transformation

$$CaCO_3(\text{calcite}) = CaCO_3(\text{aragonite})$$

$$\Delta H_m^\circ = \Delta_f H_m^\circ(\text{aragonite}) - \Delta_f H_m^\circ(\text{calcite})$$

$$= [-1207.9 - (-1207.5)]\text{kJ mol}^{-1} = -0.4 \text{ kJ mol}^{-1}$$

$$\Delta S_m^\circ = S_m^\circ(\text{aragonite}) - S_m^\circ(\text{calcite})$$

$$= (88.0 - 91.7) \text{ J mol}^{-1} \text{ K}^{-1} = -3.7 \text{ J mol}^{-1} \text{ K}^{-1}$$

$$\Delta Y_m^\circ = -\Delta H_m^\circ/T + \Delta S_m^\circ = -5.0 \text{ J mol}^{-1} \text{ K}^{-1}$$

Thus, the stable form at 298.15 K is calcite.

16.   a) If $\Delta C_{Pm,obs}^\circ$ is a constant, then

$$\Delta H_{m,obs}^\circ = \int \Delta C_{Pm,obs}^\circ \, dT = T\Delta C_{Pm,obs}^\circ + \Delta H_{m0}^\circ$$

where ΔH_{m0}° is a constant of integration. T is equal to T_H when ΔH_m° is equal to zero. Therefore

$$0 = T_H \Delta C^{\circ}_{Pm,obs} + \Delta H^{\circ}_{m,0}$$

$$\Delta H^{\circ}_{m,0} = - T_H \Delta C^{\circ}_{Pm,obs}$$

$$\Delta H^{\circ}_{m,obs} = (T - T_H)\Delta C^{\circ}_{Pm,obs}$$

$$\left(\frac{\partial(\Delta Y^{\circ}_{m,obs})}{\partial T}\right)_P = \frac{\Delta H^{\circ}_{m,obs}}{T^2} = \frac{\Delta C^{\circ}_{Pm,obs}}{T} - \frac{T_H \Delta C^{\circ}_{Pm,obs}}{T^2}$$

$$\Delta Y^{\circ}_{m,obs} = \Delta C^{\circ}_{Pm,obs} \ln T + \frac{T_H \Delta C^{\circ}_{Pm,obs}}{T} + I$$

where I is a constant of integration. Then

$$\Delta S^{\circ}_{m,obs} = \Delta Y^{\circ}_{m,obs} + \frac{\Delta H^{\circ}_{m,obs}}{T}$$

$$= \Delta C^{\circ}_{Pm,obs} \ln T + \Delta C^{\circ}_{Pm,obs} + I$$

When $\Delta S_{m,obs}{}^{\circ} = 0$, $T = T_S$, and

$$I = - \Delta C^{\circ}_{Pm,obs} (1 + \ln T_S)$$

$$\Delta Y^{\circ}_{m,obs} = \Delta C^{\circ}_{Pm,obs}\left[- \ln \frac{T_S}{T} + \frac{T_H}{T} - 1\right]$$

$$\ln K_P = \frac{\Delta Y^{\circ}_{m,obs}}{R} = \frac{\Delta C^{\circ}_{Pm,obs}}{R}\left[- \ln \frac{T_S}{T} + \frac{T_H}{T} - 1\right]$$

b) $\Delta Y_{m.obs}{}^{\circ} = R \ln K_{obs}$, and the values obtained will be tabulated below together with the others calculated. It can be seen from the derivation that the use of ΔY instead of ΔG makes the derivation easier.

c) The nonlinear fitting program that I used was unable to obtain a fit unless I had a good approximation for the 3 parameters to use as trial values. This can happen with nonlinear fitting because the program oscillates among local minima for the least squares sum of deviations and cannot find the global minimum. The trial value of T_H can be found by observing the temperature at which $\ln K_{obs}$ is a maximum, and the trial value of T_S can be found by observing the temperature at which $\Delta G_{m,obs}°$ is a minimum. The third parameter, $\Delta C_{Pm,obs}°$, is obtained by trial and error in fitting the desired function to the experimental points with trial values of T_H and T_S. With a trial value of $\Delta C_{Pm,obs}°$, the function can be fitted with only T_S and T_H as unknown values. With the values of T_S and T_H determined, a best value of $\Delta C_{Pm,obs}°$ can be obtained with the values of T_S and T_H fixed. The resulting value of $\Delta C_{Pm,obs}° = -5.08$ kJ mol^{-1} K^{-1}, $T_H = 294$ K, and $T_S = 305$ K.

The data points and fitted curve are shown in Figure 10-3, with $\ln K_{obs}$ plotted against T. Figure 10-4 shows a plot of $\Delta G_{m,obs}°$ against T, from which a trial value of T_S was obtained.

Table 10-2.

T/K	$\Delta G_{m,obs}°$/(kJ/mol)	$\Delta H_{m,obs}°$/(kJ/mol)	$\Delta S_{m,obs}°$/(J/mol/K)	$\Delta Y_{m,obs}°$/(J/mol/K)
273.15	-53.50	108.9	597.12	195.8
280.15	-58.45	71.15	460.73	208.6
287.15	-60.70	33.42	327.71	211.4
296.15	-62.82	-15.09	161.36	212.1
310.15	-64.00	-90.55	-87.60	206.4
314.15	-62.83	-112.11	-157.67	200.0

Note, that at lower temperatures $\Delta Y_{m,obs}°$ is positive because $\Delta S_{m,obs}°$ is positive, whereas at higher temperatures $\Delta Y_{m,obs}°$ is positive because $\Delta H_{m,obs}°$ is negative. Despite the large variations in $\Delta H_{m,obs}°$ and $\Delta S_{m,obs}°$, there is very little variation in $\Delta Y_{m,obs}°$. The point at 303.15 K has been omitted from consideration because it fell so far from the fitted curve.

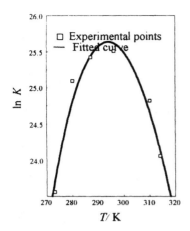

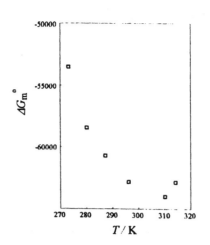

Figure 10-3. A plot of ln K_{obs} against T.   **Figure 10-4.** A plot of $\Delta G_{m,obs}°$ against T.

Chapter 11

Thermodynamics of Systems of Variable Composition

2)

$$dS \geq \frac{DQ}{T}$$

where > applies to an irreversible process and = applies to a reversible process. Nevertheless, it is correct that, under the conditions listed, an irreversible process must be spontaneous, and a reversible process must be a result of small fluctuations about a state of equilibrium.

From the first law

$$dS \geq \frac{dU}{T} - \frac{DW}{T}$$

$$\geq \frac{dU}{T} + \frac{PdV}{T}$$

when only P-V work is done and $P = P_{ex}$. That is

$$TdS \geq dU + PdV$$

a) At constant T and V, $dV = 0$, and SdT can be added to the left side to give

$$TdS + SdT \geq dU \qquad \text{or}$$

$$dU - TdS - SdT \leq 0$$

Since $\qquad dA = \Sigma \mu_i dn_i$ at constant T,V

$$\Sigma \mu_i dn_i \leq 0$$

Therefore, since $dA \leq 0$ is a criterion of equilibrium and spontaneity at constant T,V, so also is $\Sigma\mu_i dn_i$.

b) At constant S and P,

$$P = P_{ex} \, , \; VdP = 0 \, , \; TdS = 0 \, , \quad \text{and}$$

$$0 \geq dU + PdV + VdP = dH$$

$$\text{Since } dH = \Sigma\mu_{idn_i} \text{ at constant} S,P$$

$$\Sigma\mu_i dn_i \leq 0 \quad \text{at constant } S,P$$

As before, the conditions of reversibility and irreversibility are those of equilibrium and spontaneity.

c) At constant S and V, $dS = 0$ and $dV = 0$, so that

$$0 \geq dU$$

Since $dU = \Sigma\mu_i dn_i$ at constant S,V, $\Sigma\mu_i dn_i \leq 0$ at constant S,V. As before, the conditions of reversibility and irreversibility are those of equilibrium and spontaneity.

d) At constant U and V, $dU = 0$ and $dV = 0$, so that $dS \geq 0$ .Since $dS = -\Sigma(\mu_i/T)dn_i$ at constant U,V, $\Sigma(\mu_i/T)dn_i \geq 0$ at constant U,V. Since T is always positive, $\Sigma\mu_i dn_i \leq 0$ at constant U and V. Again, the conditions of reversibility and irreversibility are those of equilibrium and spontaneity.

4)   a)

$$\left(\frac{\partial G}{\partial P}\right)_{T,X_i,X_j} = V$$

If we differentiate both sides with respect to n_i at constant T,P,n_j,

$$\left[\frac{\partial}{\partial n_i}\left(\frac{\partial G}{\partial P}\right)_{T,P,X_i,X_j}\right]_{T,P,n_j} = \left(\frac{\partial V}{\partial n_i}\right)_{T,P,n_j}$$

Since G is a state function, we can reverse the order of differentiation on the left, so that

$$\left[\frac{\partial}{\partial P}\left(\frac{\partial G}{\partial n_i}\right)_{T,P,n_j}\right]_{T,X_i,X_j} = \left(\frac{\partial G_{mi}}{\partial P}\right)_{T,X_i,X_j} = V_{mi}$$

b)

$$\left(\frac{\partial G}{\partial T}\right)_{P,X_i,X_j} = -S$$

If we differentiate both sides with respect to n_i at constant T,P,n_j,

$$\left[\frac{\partial}{\partial n_i}\left(\frac{\partial G}{\partial T}\right)_{P,X_i,X_j}\right]_{T,P,n_j} = -\left(\frac{\partial S}{\partial n_i}\right)_{T,P,n_j}$$

Since G is a state function, we can reverse the order of differentiation on the left, so that

$$\left[\frac{\partial}{\partial T}\left(\frac{\partial G}{\partial n_i}\right)_{T,P,n_j}\right]_{P,X_i,X_j} = -\left(\frac{\partial G_{mi}}{\partial T}\right)_{P,X_i,X_j} = -S_{mi}$$

c)

$$G = H + T\left(\frac{\partial G}{\partial T}\right)_{P,X_i,X_j}$$

$$\left(\frac{\partial G}{\partial n_i}\right)_{T,P,n_j} = \left(\frac{\partial H}{\partial n_i}\right)_{T,P,n_j} + T\left[\frac{\partial}{\partial n_i}\left(\frac{\partial G}{\partial T}\right)_{P,X_i,X_j}\right]_{T,P,n_j}$$

$$= \left(\frac{\partial H}{\partial n_i}\right)_{T,P,n_j} + T\left[\frac{\partial}{\partial T}\left(\frac{\partial G}{\partial n_i}\right)_{T,P,n_j}\right]_{P,X_i,X_j}$$

since we can reverse the order of differentiation of a state function such as G. Thus,

$$G_{mi} = H_{mi} + T\left(\frac{\partial G_{mi}}{\partial T}\right)_{P,X_i,X_j}$$

d)

$$\left[\frac{\partial(G/T)}{\partial T}\right]_{P,X_i,X_j} = -\frac{H}{T^2}$$

$$\left\{\frac{\partial}{\partial n_i}\left[\frac{\partial(G/T)}{\partial T}\right]_{P,X_i,X_j}\right\}_{T,P,N_j} = -\frac{1}{T^2}\left(\frac{\partial H}{\partial n_i}\right)_{T,P,n_j}$$

Since G is a state function, we can reverse the order of differentiation on the left to obtain

$$\left[\frac{\partial(\mu_i/T)}{\partial T}\right]_{P,X_i,X_j} = -\frac{1}{T^2}H_{mi}$$

e)

$$\left(\frac{\partial Y}{\partial P}\right)_{T,X_i,X_j} = -\frac{V}{T}$$

$$\left\{\frac{\partial}{\partial n_i}\left(\frac{\partial Y}{\partial P}\right)_{T,X_i,X_j}\right\}_{T,P,n_j} = -\frac{1}{T}\left(\frac{\partial V}{\partial n_i}\right)_{T,P,n_j}$$

Since Y is a state function, we can reverse the order of differentiation on the left to obtain

$$\left(\frac{\partial(Y_{mi})}{\partial P}\right)_{T,X_i,X_j} = -\frac{1}{T}V_{mi}$$

6.

$$d\Delta Y_m^\circ = \left(\frac{\partial\Delta Y_m^\circ}{\partial T}\right)_{pH,pMg}dT + \left(\frac{\partial\Delta Y_m^\circ}{\partial pH}\right)_{T,pMg}d\,pH + \left(\frac{\partial\Delta Y_m^\circ}{\partial pMg}\right)_{T,pH}d\,pMg$$

$$= \frac{\Delta H_m^\circ}{T^2} dT + 2.3 R n_H d pH + 2.3 R n_{Mg} d pMg$$

Since ΔY_m° is a state function, appropriate cross derivatives are equal, so that

a)

$$2.3 R \left(\frac{\partial n_H}{\partial T} \right)_{pH,pMg} = \frac{1}{T^2} \left(\frac{\partial \Delta H_m^\circ}{\partial pH} \right)_{T,pMg}$$

$$\left(\frac{\partial \Delta H_m^\circ}{\partial pH} \right)_{T,pMg} = 2.3 R T^2 \left(\frac{\partial n_H}{\partial T} \right)_{pH,pMg}$$

b) Similarly,

$$2.3 R \left(\frac{\partial n_{Mg}}{\partial T} \right)_{pH,pMg} = \frac{1}{T^2} \left(\frac{\partial \Delta H_m^\circ}{\partial pMg} \right)_{T,pH}$$

$$\left(\frac{\partial \Delta H_m^\circ}{\partial pMg} \right)_{T,pH} = 2.3 R T^2 \left(\frac{\partial n_{Mg}}{\partial T} \right)_{pH,pMg}$$

c)

$$\left(\frac{\partial \Delta H_m^\circ}{\partial pH} \right)_{T,pMg} = 2.3 R T^2 \left(\frac{\partial n_H}{\partial T} \right)_{pH,pMg}$$

If we differentiate both sides with respect to T at constant pH, pMg, we obtain

$$\left[\frac{\partial}{\partial T} \left(\frac{\partial \Delta H_m^\circ}{\partial pH} \right)_{T,pMg} \right]_{pH,pMg} = 2.3 R T^2 \left(\frac{\partial^2 n_H}{\partial T^2} \right)_{pH,pMg} + 2(2.3) R T \left(\frac{\partial n_H}{\partial T} \right)_{pH,pMg}$$

Since ΔH is a state function, the order of differentiation is immaterial, so that,

$$\left[\frac{\partial}{\partial T}\left(\frac{\partial \Delta H^{\circ}_{m}}{\partial pH}\right)_{T,pMg}\right]_{pH,pMg} = \left[\frac{\partial}{\partial pH}\left(\frac{\partial \Delta H^{\circ}_{m}}{\partial T}\right)_{pH,pMg}\right]_{T,pMg} = \left(\frac{\partial \Delta C^{\circ}_{P,m}}{\partial pH}\right)_{T,pMg}$$

Therefore,

$$\left(\frac{\partial \Delta C^{\circ}_{Pm}}{\partial pH}\right)_{T,pMg} = 2.3R\left[T^2\left(\frac{\partial^2 n_{H}}{\partial T^2}\right)_{pH,pMg} + 2T\left(\frac{\partial n_{H}}{\partial T}\right)_{pH,pMg}\right]$$

d) By equating cross derivatives again, we obtain

$$\left(\frac{\partial n_{H}}{\partial pMg}\right)_{T,pH} = \left(\frac{\partial n_{Mg}}{\partial pH}\right)_{T,pMg}$$

8. At constant pressure, μ_1 is a function of T and X_1, so that

$$d\mu_1 = \left(\frac{\partial \mu_1}{\partial T}\right)_{X_1} dT + \left(\frac{\partial \mu_1}{\partial X_1}\right)_{T} dX_1$$

$$\left[\frac{\partial \mu_{1(\text{sat sol})}}{\partial T}\right]_P = \left(\frac{\partial \mu_1}{\partial T}\right)_{P,X_1} + \left(\frac{\partial \mu_1}{\partial X_1}\right)_{T,P}\left(\frac{\partial X_{1(\text{sat soln})}}{\partial T}\right)_P \qquad (1)$$

From the Gibbs-Duhem equation,

$n_1 d\mu_1 = - n_2 d\mu_2$, or

$X_1 d\mu_1 = - X_2 d\mu_2$, and

$$X_1\left(\frac{\partial \mu_1}{\partial X_1}\right)_{T,P} = -X_2\left(\frac{\partial \mu_2}{\partial X_1}\right)_{T,P} = X_2\left(\frac{\partial \mu_2}{\partial X_2}\right)_{T,P}$$

$$\left(\frac{\partial \mu_1}{\partial X_1}\right)_{T,P} = \frac{X_2}{X_1}\left(\frac{\partial \mu_2}{\partial X_2}\right)_{T,P} \qquad (2)$$

At equilibrium, $\mu_{2(\text{solid})} = \mu_{2(\text{sat soln})}$, and

$$d\mu_{2(\text{solid})} = d\mu_{2(\text{sat soln})}$$

$\mu_{2(\text{solid})}$ is a function of T only, whereas $\mu_{2(\text{sat soln})}$ is a function of T and X_2. Thus,

$$\left(\frac{\partial \mu_{2(solid)}}{\partial T}\right)_P dT = \left(\frac{\partial \mu_{2(sat\ soln)}}{\partial T}\right)_{P,X_2} dT + \left(\frac{\partial \mu_{2(sat\ soln)}}{\partial X_2}\right)_{T,P} dX_2$$

Since $\quad \left(\frac{\partial \mu}{\partial T}\right)_{P,X} = S_m$

$$- S_{m2(solid)} dT = - S_{m2(sat\ soln)} dT + \left(\frac{\partial \mu_{2(sat\ soln)}}{\partial X_2}\right)_{T,P} dX_2$$

$$\left(\frac{\partial X_{2(sat\ soln)}}{\partial T}\right)_P = \frac{S_{m2(sat\ soln)} - S_{m2(solid)}}{\left(\frac{\partial \mu_2}{\partial X_2}\right)_{T,P}} \qquad (3)$$

If we substitute from Equation (2) into Equation (1), we obtain

$$\left(\frac{\partial \mu_{1(sat\ soln)}}{\partial T}\right)_P = - S_{m1} + \frac{X_2}{X_1}\left(\frac{\partial \mu_2}{\partial X_2}\right)_{T,P}\left(\frac{\partial X_{1(sat\ soln)}}{\partial T}\right)_P$$

$$= - S_{m1} - \frac{X_2}{X_1}\left(\frac{\partial \mu_2}{\partial X_2}\right)_{T,P}\left(\frac{\partial X_{2(sat\ soln)}}{\partial T}\right)_P \qquad (4)$$

If we now substitute from Equation (3) into Equation (4), we obtain

$$\left(\frac{\partial \mu_{1(satsoln)}}{\partial T}\right)_P = - S_{m1} - \frac{X_2}{X_1}[S_{m2} - S_{m2(solid)}]$$

Chapter 12

Mixtures of Gases

2. For He at 0.1 MPa and 60°C, $B = 11.5$ cm^3 mol^{-1}.

g,h)

$$\ln \gamma = \ln (f/P) = \frac{BP}{RT}$$

$$= \frac{(11.5 \times 10^{-6} \text{ m}^3 \text{ mol}^{-1})(0.1 \times 10^6 \text{ Pa})}{(8.3145 \text{ J mol}^{-1} \text{ K}^{-1})(333.15 \text{ K})} = 4.15 \times 10^{-4}$$

$\gamma = 1.0004 = f/P$

$f = (1.0004)(0.1 \times 10^6 \text{ Pa}) = 1.0004 \times 10^5$ Pa ; $\ln f = 11.513$

i) $H_m^* - H_m = -BP = (11.5 \times 10^{-6} \text{ m}^3 \text{ mol}^{-1})(0.1 \times 10^6 \text{ Pa}) = -1.15$ J mol^{-1}

j) $\mu_{J.T.} = -(B/C_{Pm}) = -(11.5 \times 10^{-6} \text{ m}^3 \text{ mol}^{-1})/[(5/2)(8.3145 \text{ J mol}^{-1} \text{ K}^{-1})] = 5.53 \times 10^{-7}$ K Pa^{-1}

4.   a) Table 12-1 shows the calculations for the numerical integration to obtain the area under the α against P curve for H$_2$, which is shown in Figure 12-1.

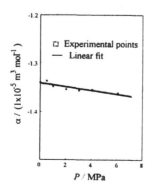

Figure 12-1. A plot of α against P for hydrogen at 250 K.

The value of α was calculated from the relation

$$\alpha = \frac{RT}{P} - V_m = \frac{RT}{P} - \frac{PV_m}{RT}\frac{RT}{P} = \frac{RT}{P}(1 - z)$$

where z is equal to $(PV_m)/(RT)$, as reported by Johnson and White. The fourth column is equal to the integral of αdP from $P = 0$ to the value of P on that line. We can use linear interpolation to obtain a value of - 67.717 for the integral at 5 MPa, which is equal to - $RT \ln (f/P)$. Thus

$$\ln (f/P) = \frac{67.717 \text{ J mol}^{-1}}{(8.3145 \text{ J mol}^{-1} \text{ K}^{-1})(250K)} = .03258$$

$(f/P) = 1.0331$, and

$f = (1.0331)(5 \text{ MPa}) = 5.166 \text{ MPa}$

Table 12-1. Calculation of f/P by the α method.

P/MPa	z	$\alpha/(10^{-5} \text{ m}^3)$	ΔP	$\ddot{\alpha}/(10^{-5} \text{ m}^3)$	$\ddot{\alpha}\Delta P$	$\Sigma\ddot{\alpha}\Delta P$	$\ln f/P$	f/P
0.000000	1.000000	-1.3400				0	0	1
0.101325	1.000674	-1.3827	0.101325	-1.3613	-1.40	-1.4010	0.000674	1.000674
0.202650	1.001308	-1.3416	0.101325	-1.3622	-1.36	-2.7604	0.001328	1.001329
0.506625	1.003258	-1.3367	0.303975	-1.3392	-4.06	-6.8237	0.003283	1.003288
1.013250	1.006572	-1.3482	0.506625	-1.3425	-6.83	-13.654	0.006569	1.00659
2.026500	1.013202	-1.3542	1.01325	-1.3512	-13.70	27.375	0.01317	1.013257
3.039750	1.019832	-1.3561	1.01325	-1.3551	-13.70	-41.116	0.01978	1.019977
4.053000	1.026412	-1.3546	1.01325	-1.3554	-13.70	-54.841	0.026383	1.026735

P/MPa	z	$\alpha/(10^{-5}\text{ m}^3)$	ΔP	$\ddot\alpha/(10^{-5}\text{ m}^3)$	$\ddot\alpha\Delta P$	$\Sigma\ddot\alpha\Delta P$	$\ln f/P$	f/P
5.066250	1.033139	-1.3597	1.01325	-1.3571	-13.80	-68.618	0.033011	1.033562
6.079500	1.039818	-1.3614	1.01325	-1.3605	-13.80	-82.412	0.039648	1.040444
7.092750	1.046691	-1.3683	1.01325	-1.3649	-13.90	-96.277	0.046318	1.047407

b) Table 12-2 is the spreadsheet for calculating f/P from the Redlich-Kwong equation and the data of Johnson and White. The values of a and b are calculated from Equations 5-58, with values of T_c and P_c from Table 5-3. $a = 0.144374$; $b = 1.8389\times10^{-5}$.

Table 12-2. Spreadsheet for calculation of f/P with the Redlich-Kwong equation.

P/MPa	z	α	$\ln f/P$	f/P	f
0.000000	1.000000		0.000000	1.000000	0.000000
0.101325	1.000674	0.020528260	0.000872	1.000872	0.101413
0.202650	1.001308	0.010270633	0.001605	1.001606	0.202975
0.506625	1.003258	0.004116254	0.003765	1.003772	0.508536
1.013250	1.006572	0.002064925	0.007334	1.007361	1.020708
2.026500	1.013202	0.001039263	0.014586	1.014693	2.056276
3.039750	1.019832	0.000697376	0.021992	1.022236	3.107342
4.053000	1.026412	0.000526407	0.029599	1.030042	4.174759
5.066250	1.033139	0.000423885	0.0372	1.037901	5.258264

P/MPa	z	α	$\ln f/P$	f/P	f
6.079500	1.039818	0.000355521	0.044992	1.046019	6.359275
7.092750	1.046691	0.000306747	0.052717	1.054132	7.476692

The values of f calculated from the Redlich-Kwong equation are very close to those calculated from th

α method.

6.   a) $N_2(g) + C_2H_2(g) = 2\, HCN(g)$

The WebBook provides data for $\Delta_f H_m^\circ$ and S_m°, but not $\Delta_f G_m^\circ$ or $\Delta_f Y_m^\circ$. Therefore, we shall calcula

ΔH_m° and ΔS_m°, from which we can calculate ΔY_m°.

$$\Delta H_m^\circ = 2\,\Delta_f H_m^\circ (HCN) - \Delta_f H_m^\circ (C_2H_2)$$

$$= 2\,(135.1432 \text{ kJ mol}^{-1}) - 226.7314 \text{ kJ mol}^{-1} = 43.555 \text{ kJ mol}^{-1}$$

$$\Delta S_m^\circ = 2\,S_m^\circ (HCN) - S_m^\circ (N_2) - S_m^\circ (C_2H_2)$$

$$= 2(201.82 \text{ J mol}^{-1} \text{ K}^{-1}) - 191.56 \text{ J mol}^{-1} \text{ K}^{-1} - 200.93 \text{ J mol}^{-1} \text{ K}^{-1}$$

$$= 11.15 \text{ J mol}^{-1} \text{ K}^{-1}$$

$$\Delta Y_m^\circ = \Delta S_m^\circ - \frac{\Delta H_m^\circ}{T} = 11.15 \text{ J mol}^{-1} \text{ K}^{-1} - \frac{43566 \text{ J mol}^{-1}}{298.15 \text{ K}} = -134.93 \text{ J mol}^{-1} \text{ K}^{-1}$$

b) In order to calculate ΔY_m° at 300°C, we need to know ΔH_m° at 300°C and ΔS_m° at 300°C. We sha

use the Shomate equations of the WebBook, which represent C_{Pm}° as polynomials in t, where t

$(T/K)/1000$.

$$\Delta H_m^\circ (573.15 \text{ K}) = \Delta H_m^\circ (298.15 \text{ K}) + \int_{273.15 \text{ K}}^{573.15 \text{ K}} \Delta C_{Pm}^\circ dT \qquad \text{where}$$

$$\Delta C_{Pm} = \Delta A + \Delta Bt + \Delta Ct^2 + \Delta Dt^3 + \frac{\Delta E}{t^2} \qquad \text{and}$$

$$\Delta S_m^{\circ}(573.15 \text{ K}) = \Delta S_m^{\circ}(298.15 \text{ K}) + \int_{273.15 \text{ K}}^{573.15 \text{ K}} \frac{\Delta C_{Pm}^{\circ}}{T} dT$$

The results are

$$\Delta H_m^{\circ} = 43.1918 \text{ kJ mol}^{-1}$$

$$\Delta S_m^{\circ} = 10.93 \text{ J mol}^{-1} \text{ K}^{-1} \qquad \text{and}$$

$$\Delta Y_m^{\circ} = -64.43 \text{ J mol}^{-1} \text{ K}^{-1}$$

c) $\qquad \ln K_f = \Delta Y_m^{\circ}/R$

At 298.15 K, $\ln K_f = (-134.93 \text{ J mol}^{-1} \text{ K}^{-1})/(8.3145 \text{ J mol}^{-1} \text{ K}^{-1}) = -16.228$

$$K_f = 8.96 \times 10^{-8}$$

At 573.15 K, $\ln K_f = (-64.63 \text{ J mol}^{-1} \text{ K}^{-1})/(8.3145 \text{ J mol}^{-1} \text{ K}^{-1}) = -7.773$

$$K_f = 4.21 \times 10^{-4}$$

The variation of K_f with temperature is consistent with Le Chatelier's principle.

Values of thermodynamic properties for HCN from earlier tables (which differ by 20-25%) lead to values

of K_f equal to 2.12×10^{-6} at 298.15 K and 4.38×10^{-3} at 573.15 K. The differences in the calculated values

for K_f for relatively small differences in ΔY_m° and ΔH_m° emphasize the need for precision and accuracy

in thermodynamic functions used to calculate equilibrium constants, since K_f is an exponential function

of ΔY_m°.

d) Table 12-3, taken from a spreadsheet indicates the calculation of the reduced variables from the experimental

values and the critical comstants.

Table 12-3.

	T_C /K	P_C /MPa	T_r	P_r 0.5 MPA	P_r 20.0 MPa
N_2	126.0	3.39	4.55	0.147	5.90
C_2H_2	309.0	6.25	1.855	0.080	3.20
HCN	457.4	5.07	1.253	0.099	3.94

e),f),g) Table 12-4 shows the values of activity coefficients obtained from Figure 12-5 and the calculat

values of the equilibrium constants.

Table 12-4.

P/MPa	$\gamma(N_2)$	$\gamma(C_2H_2)$	$\gamma(HCN)$	K_γ	K_f /10^{-4}	K_P /10^{-4}
0.5	1.00	1.00	0.98	0.96	3.99	4.16
20.0	1.1	0.88	0.53	0.29	3.99	13.8

$K_\gamma = [\gamma^2(HCN)]/[\gamma(N_2)\gamma(C_2H_2)]$, with the results above.

$K_P = K_f /K_\gamma$, with the results above.

h)

	$N_2(g)$ +	$C_2H_2(g)$ =	2 HCN(G)	total
n	$1 - \alpha$	$1 - \alpha$	2α	$1 + \alpha$
X	$(1 - \alpha)/(1 + \alpha)$	$(1 - \alpha)/(1 + \alpha)$	$2\alpha/(1 - \alpha)$	
p	$(1 - \alpha)P/(1 + \alpha)$	$(1 - \alpha)P/(1 + \alpha)$	$2\alpha P/(1 - \alpha)$	

$$K_P = \frac{\left[\left(\frac{2\alpha}{1+\alpha}\right)P\right]^2}{\left[\left(\frac{1-\alpha}{1+\alpha}\right)P\right]^2} = \frac{4\alpha^2}{(1-\alpha)^2}$$

At 0.5 MPA

$$\frac{4\alpha^2}{(1-\alpha)^2} = 4.16\times10^{-4} \quad ; \quad \frac{2\alpha}{1-\alpha} = 2.04\times10^{-2} \quad ; \quad \alpha = 1.03\times10^{-2}$$

At 20.0 MPa

$$\frac{4\alpha^2}{(1-\alpha)^2} = 1.38\times10^{-3} \quad ; \quad \frac{2\alpha}{1-\alpha} = 3.71\times10^{-2} \quad ; \quad \alpha = 1.89\times10^{-2}$$

i) An increase in pressure increases the yield of HCN.

j) According to Le Chatelier's principle, the equilibrium yield should be independent of pressure since $\Delta n = 0$, but that principle does not take non-ideality into account.

a) Table 12-5 from a spreadsheet illustrates the calculation of $\alpha = (RT/P) - V_m$ from data on the molar volume of O_2 as a function of P.

Table 12-5. Molar volume and α for Oxygen at 300 K as a function of pressure.

P/MPa	V_m/(cm^3 mol^{-1})	z	α/(cm^3 mol^{-1})
0.0000		1.00000	16.243
4.6800	518.70	0.96256	14.292
7.1046	337.95	0.96256	13.144
7.8823	303.82	0.96008	12.634
9.2470	257.84	0.95586	11.907
10.411	228.17	0.95232	11.423

P/MPa	V_m/(cm^3 mol^{-1})	z	α/(cm^3 mol^{-1})
11.901	198.88	0.94884	10.722
12.769	185.01	0.94709	10.336
14.539	162.22	0.94555	9.342
15.495	152.10	0.94482	8.882
15.786	149.33	0.94504	8.685
19.119	123.70	0.94819	6.772
21.988	108.20	0.95379	5.242
24.8740	96.65	0.96374	3.636
27.0831	89.69	0.97382	2.412
28.254	86.50	0.97981	1.782
29.918	82.46	0.98902	0.916

Since the data are highly precise, it is essential to use a precise value for R, 8.31451 J mol^{-1} K^{-1}. The curve for α as a function of P is shown in Figure 12-2, together with a fitted cubic polynomial.

$$\alpha = 16.2432 - 0.4016P - 0.00648P^2 + 0.00009166P^3$$

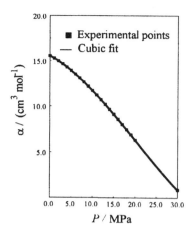

Figure 12-2. Alpha as a function of P for oxygen at 300 K. Data from H. L. Johnston and D. White, *Trans.*

Am. Soc. Mech. Engineers **72**, 785 (1950).

b,c)

$$\ln \frac{f}{P} = -\frac{1}{RT}\int_0^P \alpha \, dP$$

$$= -\frac{1}{RT}\int_0^P (16.2432 - 0.4016P - 0.00648P^2 + 9.1659\times10^{-5}P^3)\,dT$$

$$= -\frac{1}{RT}(16.2432P - 0.2008P^2 - 0.00216P^3 + 2.2915\times10^{-5}P^4)$$

$$= -6.5120\times10^{-3}P + 8.051\times10^{-5}P^2 + 8.660\times10^{-7}P^3 - 9.187\times10^{-9}P^4$$

The values of f/P calculated from values of α and the corresponding values from the approximation that

$f/P = z$ are shown in Table 12-6.

Table 12-6. Values of ln f/P calculated from Equation 12-46 by analytical integration, and from Equation 12-65

P/MPa	z	ln f/P	ln z	f/P
4.680	0.973185	-0.02871	-0.02718	0.971691
7.105	0.962562	-0.04213	-0.03815	0.958739
7.882	0.960077	-0.04605	-0.04074	0.954992
9.247	0.955858	-0.05297	-0.04514	0.948403
10.411	0.952324	-0.05849	-0.04855	0.943184
11.901	0.948845	-0.06530	-0.05250	0.936785
12.769	0.947088	-0.06880	-0.05436	0.933504
14.539	0.945548	-0.07626	-0.05599	0.926567
15.495	0.944825	-0.07944	-0.05675	0.923630
15.786	0.945039	-0.08008	-0.05652	0.923039
19.119	0.948096	-0.09289	-0.05329	0.911289
21.988	0.953789	-0.10035	-0.04631	0.904511
24.874	0.963742	-0.10647	-0.03693	0.898993
28.254	0.979811	-0.10837	-0.02039	0.897293
29.918	0.989019	-0.11059	-0.01104	0.895300

d) We can see that the approximation is good to about 1% for f/P for pressures below 11 Mpa, even though α varies by approximately 20% in that pressure range. The agreement is not nearly as good for $\ln z$ and $\ln f/P$. If the value of α were constant at its value at 10.4 MPa, the value of $\ln (f/P)$ would equal -0.04855, and f/P would equal 0.95323. With the exact method $\ln (f/P)$ equals -0.05849 and f/P equals 0.943184. Even though the natural logs differ by more than 20%, the arguments of the natural logs vary only by 1%. Small differences in the argument of the natural log in this range result in large differences in the natural logarithm. Remember that $[(d \ln x)/dx] = 1/x$, so that at small x the change in ln is large for a small change in x.

Chapter 13

The Phase Rule

2.

$$\frac{dP}{dT} = \frac{\Delta H_m}{T \Delta V_m}$$

$$= \frac{728 \pm 167 \text{ J mol}^{-1}}{(847 \pm 1.5 \text{ K})(0.154 \pm 0.015 \times 10^{-6} \text{ m}^3 \text{ mol}^{-1})}$$

$$= 5.58 \pm 1.28 \times 10^6 \text{ J m}^{-3} \text{ K}^{-1}$$

$$= 5.58 \pm 1.28 \times 10^6 \text{ Pa K}^{-1}$$

$$\frac{dT}{dP} = \frac{1}{5.58 \pm 1.28 \text{ MPa K}^{-1}} = 0.18 \pm 0.04 \text{ K/MPa}$$

which is indistinguishable experimentally from the value of 0.21 K/MPa of R. E. Gibson.

4. $F = C - \mathcal{P} + 1$, for a two component system at fixed pressure. If two phases are present at equilibrium, $F = 1$, and the state of the system must be represented by curves, which have one degree of freedom, one for each phase. If one phase is present at equilibrium, $F = 2$, and the state of the system must be represented by an area, which has two degrees of freedom, since both variables can be varied freely within limits. One can recognize a curve in the phase rule sense by noting that it must represent a function, which is continuous and has a continuous first derivative in the range of temperatures in which the phases are stable .

Region I. Since the right boundary has a discontinuous slope, this region represents an area with two degrees of freedom, and must represent one phase, which is a solid solution of Na and Na$_2$K.

Region II. The right and left boundary curves have a continuous first derivative and are functions with one degree of freedom; thus the region represents two phases, solid solution I in equilibrium with liquid VIII. Curve *CB* represents the composition of the solid solution. and curve *CD* represents the composition of liquid VIII.

Region III. Bounded by curves; two phases, solid solution I and solid Na_2K. Curve *BA* represents the composition of solid solution I; curve *EF* represents the constant composition of Na_2K.

Region IV. Bounded by curves; two phases, solid Na_2K and liquid VIII. Curve *DH* represents the composition of solution VIII; curve *EG* represents the constant composition of Na_2K.

Region V. Bounded by curves; two phases, solid Na_2K and solid solution VII. Curve *MN* represents the composition of solution VII; curve *GF* represents the constant composition of Na_2K.

Region VI. Bounded by curves; two phases, solid solution VII and liquid solution VIII. Curve *GJ* represents the composition of liquid VIII; curve *MJ* represents the composition of solution VII.

Region VII. The left boundary has discontinuous slope; thus, one phase, solid solution VII.

Region VIII. Boundaries have discontinuous slopes; thus, one phase, liquid solution VIII.

Chapter 14

The Ideal Solution

2.

$$\Delta S_m = -\frac{1}{T}[G_m(\text{components}) - G_m(\text{mixture})]$$

$$= -\frac{1}{T}[X_1 \mu_1^\bullet + X_2 \mu_2^\bullet - \{X_1(\mu_1^\bullet + RT \ln X_1) + X_2(\mu_2^\bullet + RT \ln X_2)\}]$$

$$= R(X_1 \ln X_1 + X_2 \ln X_2)$$

$$= 8.3145 \text{ J mol}^{-1} \text{ K}^{-1}[0.007 \ln (0.007) + 0.993 \ln (0.993)]$$

$$= -0.347 \text{ J mol}^{-1} \text{ K}^{-1}$$

4.   If the gas forms an ideal solution in the liquid,

$$\mu_2 = \mu_2^\circ(\text{superheated liquid}) + RT \ln X_2$$

and, at equilibrium of gas with saturated solution,

$$\mu_2(\text{satd}) = \mu_2^\circ(\text{superheated liquid}) + RT \ln X_2(\text{satd})$$

$$= \mu_2^\bullet(\text{gas}) \quad or$$

$$RT \ln X_2(\text{satd}) = \mu_2^\bullet(\text{gas}) - \mu_2^\bullet(\text{superheated liquid})$$

$$\ln X_2(\text{satd}) = \frac{1}{RT}[\mu_2^\bullet(\text{gas}) - \mu_2^\bullet(\text{superheated liquid})]$$

a)

$$\left(\frac{\partial \ln X_{2,\text{satd}}}{\partial P}\right)_T = \frac{1}{RT}\left[\left(\frac{\partial \mu_2^\bullet(\text{gas})}{\partial P}\right)_T - \left(\frac{\partial \mu_2^\bullet(\text{superheated liquid})}{\partial P}\right)_T\right]$$

$$= \frac{1}{RT}\left(V_{m2.g} - V_{m2}\right)$$

since $V_{m2} = V_{m2}^\bullet$ for an ideal solution.

b)

$$\ln X_2(\text{satd}) = \frac{1}{R}\left[\frac{\mu_2^\bullet(\text{gas})}{T} - \frac{\mu_2^\bullet(\text{superheated liquid})}{T}\right]$$

$$\left(\frac{\partial \mu_{2,\text{satd}}}{\partial T}\right)_P = \frac{1}{R}\left\{\left[\frac{\partial(\mu_2^\bullet(\text{gas})/T)}{\partial T}\right]_P - \left[\frac{\partial(\mu_2^\bullet(\text{superheated liquid})/T)}{\partial T}\right]_P\right\}$$

$$= \frac{1}{R}\left[-\frac{H_{m2}^\bullet(\text{gas})}{T^2} + \frac{H_{m2}^\bullet(\text{superheated liquid})}{T^2}\right]$$

$$= -\frac{\Delta H_{m2,\text{vap}}}{RT^2}$$

c) When pure liquid solute at the boiling point (no longer supercooled) is mixed with the solvent, there is no limit to the solubility, since the solution is ideal. Therefore we can integrate the last equation in b) from $X_2 = 1$ at T_b to X_2 at some temperature T.

$$\int_{X_2 = 1}^{X_2(\text{satd})} d \ln X_2(\text{satd}) = \int_{T_b}^T - \frac{\Delta H_{m2,\text{vap}}}{RT^2} dT$$

If $\Delta H_{m2,\text{vap}}$ is constant, the result is

$$\ln X_2 = -\frac{\Delta H_{m2,vap}}{R}\left(-\frac{1}{T} + \frac{1}{T_b}\right)$$

This result is consistent with the initial assumption that X_2(satd) is equal to 1 at T_b, and shows

that X_2(satd) decreases continuously with T above T_b.

6.   Equilibrium between the isomers is attained when X_3 is equal to 0.372. If the solution is ideal,

$$\Delta G_m = \Delta G_m^\circ + RT \ln\left(\frac{X_2}{X_3}\right)_{\text{equilibrium}}$$

$$\ln\left(\frac{X_2}{X_3}\right)_{\text{equilibrium}} = -\frac{\Delta G_m^\circ}{RT} = -\frac{(-1194\ \text{J mol}^{-1})}{(8.3145\ \text{J mol}^{-1}\ \text{K}^{-1})(273.15\text{K})}$$

$$= 0.528$$

$$\left(\frac{X_2}{X_3}\right) = 1.692$$

$$\left(\frac{1-X_3}{X_3}\right) = 1.692, \quad X_{3,\text{equilibrium}} = \frac{1}{2.692} = 0.371$$

in agreement with the experimental value.

Chapter 15

Dilute Solutions of Nonelectrolytes

2.   If a solute dissociates completely into two particles in dilute solution, Henry's law is expressed as

$$f_2 \text{ (gas)} = K (X_{2,cond})^2$$

where X_2 is calculated as if there were no dissociation.

At equilibrium between the dissolved solute and its gas phase

$$\mu_{2,cond} = \mu_{2,gas} = \mu_{2,gas}^\circ + RT \ln (f_2/f^\circ)$$

$$= \mu_{2,gas}^\circ + RT \ln(K X_2^2/f^\circ)$$

$$= \mu_{2,gas}^\circ + RT \ln \frac{K}{f^\circ} + RT \ln X_2^2$$

$$= \mu_{2,cond}^\circ + RT \ln X_2^2$$

where $\mu_{2,cond}^\circ \equiv \mu_{2,gas}^\circ + RT \ln (K/f^\circ)$ and equals the chemical potential of solute in the condensed phase extrapolated to $X_2 = 1$ along the Henry's law line of the dilute solution.

a) If the solute dissociates in solvent I and not in solvent II and X_2 is calculated as before as if there were no dissociation,

$$\mu_2(I) = \mu_2^\circ(I) + RT \ln X_2^2(I)$$

$$\mu_2(II) = \mu_2^\circ(II) + RT \ln X_2(II)$$

When the solute is distributed at equilibrium between the two solvents, the two chemical potentials are equal, and

$$\mu_2^{\circ}(I) + RT \ln X_2^2(I) = \mu_2^{\circ}(II) + RT \ln X_2(II)$$

$$RT \ln \frac{X_2^2(I)}{X_2(II)} = \mu_2^{\circ}(II) - \mu_2^{\circ}(I)$$

$$\frac{X_2^2(I)}{X_2(II)} = e^{\frac{\mu_2^{\circ}(II) - \mu_2^{\circ}(I)}{RT}} = \kappa$$

where κ is the Nernst distribution constant for the solute in the two solvent systems.

b) In the osmotic pressure experiment

$$d\mu_1^{\bullet} = d\mu_1 = 0$$

$$= \left(\frac{\partial \mu_1}{\partial P}\right)_{T,X_2} dP + \left(\frac{\partial \mu_1}{\partial X_2}\right)_{T,P} dX_2$$

$$\left(\frac{\partial \mu_1}{\partial P}\right)_{T,X_2} = V_{m1} \qquad \text{and}$$

$$0 = V_{m1} dP + \left(\frac{\partial \mu_1}{\partial X_2}\right)_{T,P}$$

To find $(\partial \mu / \partial X_2)_{T,P}$, we use the Gibbs-Duhem equation,

$$X_1 d\mu_1 + X_2 d\mu_2 = 0 \qquad \text{and}$$

$$X_1 \left(\frac{\partial \mu_1}{\partial X_2}\right)_{T,P} + X_2 \left(\frac{\partial \mu_2}{\partial X_2}\right)_{T,P} = 0$$

Therefore,

$$\left(\frac{\partial \mu_1}{\partial X_2}\right)_{T,P} = -\frac{X_2}{X_1}\left(\frac{\partial \mu_2}{\partial X_2}\right)_{T,P}$$

$$= -\frac{X_2}{X_1}\frac{RT}{X_2^2}(2X_2) = -\frac{2RT}{X_1}$$

Then,

$$0 = V_{m1}\,dP - \frac{2RT}{X_1}\,dX_2$$

$$\frac{dP}{dX_2} = \frac{2RT}{X_1 V_{m1}}$$

Since, the solvent follows Raoult's law, $V_{m1} = V_{m1}^{\bullet}$, and

$$\frac{dP}{dX_2} = \frac{2RT}{V_{m1}^{\bullet}}$$

$$dP = \frac{2RT}{V_{m1}^{\bullet}}\,dX_2$$

$$\int_{P_0}^{P} dP = \int_{0}^{X_2}\frac{2RT}{V_{m1}^{\bullet}}\,dX_2$$

$$\Pi = P - P_0 = \frac{2RT}{V_{m1}^{\bullet}}X_2$$

In solution sufficiently dilute to follow Henry's law, $X_2 \approx n_2/n_1$, and

$$\Pi = \frac{2RTn_2}{n_1 V_{m1}^{\bullet}} \approx \frac{2RTn_2}{V} = 2RTC_2$$

where n_2 is calculated as if there were no dissociation. Clearly, the osmotic pressure is twice what

it would be if there were no dissociation. A similar modification of the derivation for freezing

point depression and boiling point elevation leads to

$$\Delta T_F = 2 K_F m_2$$

$$\Delta T_B = 2 K_B m_2$$

4.   a) $(20 \text{ g kg}^{-1} / 180.16 \text{ g mol}^{-1}) = 0.111 \text{ mol kg}^{-1}$

The transformation for which we want to calculate $\Delta_f G_m°$ is

6 C(graphite) + 6 H_2(g) + 3 O_2(g) = α-$C_6H_{12}O_6$(1.00 mol kg^{-1}, hypothetical)

We can obtain that transformation as the sum of the following steps:

1.   6 C(graphite) + 6 H_2(g) + 3 O_2(g) = α-$C_6H_{12}O_6$(s)

$$\Delta G_{m1} = \Delta_f G_m° = -902900 \text{ J mol}^{-1}$$

2. α-$C_6H_{12}O_6$(s) = α-$C_6H_{12}O_6$(0.111 mol kg^{-1})

$$\Delta G_{m2} = 0$$

3. α-$C_6H_{12}O_6$(0.111 mol kg^{-1}) = α-$C_6H_{12}O_6$(1.00 mol kg^{-1}, hypothetical)

$$\Delta G_{m3} = RT \ln [(1.00 \text{ mol kg}^{-1})/(0.111 \text{ mol kg}^{-1})] = 5446 \text{ mol kg}^{-1}$$

The desired value is

$$\Delta G_m = \Delta_f G_m°(\text{dissolved } \alpha) = \Delta G_{m1} + \Delta G_{m2} + \Delta G_{m3}$$

$$= -897500 \text{ J mol}^{-1}$$

b) $(49 \text{ g kg}^{-1} / 180.16 \text{ g mol}^{-1}) = 0.272 \text{ mol kg}^{-1}$

$$\Delta_f G_m°(\text{dissolved } \beta) = \Delta_f G_m°(\text{solid } \beta) + 0$$

$$+ RT \ln [(1.00 \text{ mol kg}^{-1})/(0.272 \text{ mol kg}^{-1})]$$

$$= -898000 \text{ J mol}^{-1}$$

For the transformation

$$\alpha\text{-}D\text{-glucose} = \beta\text{-}D\text{-glucose}$$

$$\Delta G_m^\circ = \Delta_f G_m^\circ(\beta) - \Delta_f G_m^\circ(\alpha)$$

$$= [- 898000 - (-897500)] \text{ J mol}^{-1} = 500 \text{ J mol}^{-1}$$

$$\ln K = - \Delta G_m^\circ / (RT) = 0.202$$

$$K = 1.22$$

6.   The transformation for which we wish to calculate ΔG_m° is

$$O_2(g, 100 \text{ kPa}) = O_2[\text{aqueous}, m2 = 1 \text{ (hypothetical)}]$$

We can express that transformation as the sum of the following steps:

1. $O_2(g, 100 \text{ kPa}) = O_2(\text{air}, 20.26 \text{ kPa})$ (air is 20.26 mol % O2)

$$\Delta G_{m1} = RT \ln [(20.26 \text{ kPa})/(100 \text{ kPa})] = - 3955 \text{ J mol-1}$$

2. $O_2(\text{air}, 20.26 \text{ kPa}) = O_2(\text{aqueous}, 0.0023 \text{ mol kg}^{-1})$

$$\Delta G_{m2} = 0$$

3. $O_2(\text{aqueous}, 0.0023 \text{ mol kg}^{-1}) = O_2[\text{aqueous}, m_2 = 1 \text{ (hypothetical)}]$

$$\Delta G_{m3} = RT \ln [(1.00 \text{ mol kg}^{-1})/(0.0023 \text{ mol kg}^{-1})] = 20,800 \text{ J mol}^{-1}$$

$$\Delta_f G_m^\circ [O2(\text{aqueous})] = \Delta G_{m1} + \Delta G_{m2} + \Delta G_{m3} = 16.8 \text{ kJ mol-1}$$

8.   The transformation for which we wish to calculate $\Delta_f G_m^\circ$ is

$$N_2(g, 100 \text{ kPa}) = N_2[\text{aqueous}, m_2 = 1.00(\text{hypothetical})]$$

This transformation can be expressed as the sum of the following transformations:

1. $N_2(g, 100 \text{ kPa}) = N_2(\text{air}, 81 \text{ kPa})$

$$\Delta G_{m1} = RT \ln [(81 \text{ kPa})/(100 \text{ kPa})] = - 478 \text{ J mol-1}$$

2.
$$N_2(\text{air, 81 kPa}) = N_2(\text{aqueous, } 0.84 \times 10^{-3} \text{ mol kg}^{-1})$$

$$\Delta G_{m2} = 0$$

3. $N_2(\text{aqueous, } 0.84 \times 10^{-3} \text{ mol kg}^{-1}) = N_2[\text{aqueous, } m_2 = 1.00(\text{hypothetical})]$

$$\Delta G_{m3} = RT \ln [(1.00)/(0.84 \times 10^{-3})] = 16,080 \text{ kJ mol}^{-1}$$

$$\Delta_f G_m^\circ = (16,080 - 478) \text{ J mol-1} = 15.60 \text{ kJ mol}^{-1}$$

Chapter 16

Activities, Excess Gibbs Functions, and Standard States for Nonelectrolytes

2.   a)

$$RT^2\left(\frac{\partial \ln a_2}{\partial T}\right)_{P,M_2} = H^\circ_{m2} - H_{m2}$$

b)

$$RT^2\left(\frac{\partial^2 \ln a_2}{\partial T^2}\right)_{P,m_2} + 2RT\left(\frac{\partial \ln a_2}{\partial T}\right)_{P,m_2} = \left(\frac{\partial H_{m2}}{\partial T}\right)_{P,m_2} + \left(\frac{\partial H^\circ_{m2}}{\partial T}\right)_{P,m_2}$$

c) From the definition of the activity coefficient

$$\left(\frac{\partial a_2}{\partial T}\right)_{P,m_2} = \left[\frac{\partial \ln (\gamma_2 m_2/m^\circ_2)}{\partial T}\right]_{P,m_2} = \left(\frac{\partial \ln \gamma_2}{\partial T}\right)_{P,m_2}$$

and

$$\left(\frac{\partial^2 \ln a_2}{\partial T}\right)_{P,m_2} = \left(\frac{\partial^2 \ln \gamma_2}{\partial T^2}\right)_{P,m_2}$$

In the infinitely dilute solution, $\gamma_2 = 1$, and

$$\left(\frac{\partial \ln \gamma_2}{\partial T}\right)_{P,m_2} = \left(\frac{\partial^2 \ln \gamma_2}{\partial T^2}\right)_{P,m_2} = 0$$

so that,

$$0 = \left(\frac{\partial H^{\circ}_{m2}}{\partial T}\right)_{P,m_2} - \left(\frac{\partial H^{\infty}_{m2}}{\partial T}\right)_{P.m_2}$$

$$= C^{\circ}_{Pm2} - C^{\infty}_{Pm2}$$

$$C^{\infty}_{Pm2} = C^{\circ}_{Pm2}$$

4.   For any solution, the chemical potential is independent of the scale used, so that

$$\mu^{\circ}_C + RT \ln a_C = \mu^{\circ}_m + RT \ln a_m$$

$$\ln\frac{a_m}{a_C} = \frac{\mu^{\circ}_C - \mu^{\circ}_m}{RT}$$

$$= \frac{\mu^{\circ}_{2,gas} + RT \ln (k_2'''/f^{\circ}) - \mu^{\circ}_{2,gas} - RT \ln (k_2''/f^{\circ})}{RT}$$

$$= \ln(k_2'''/k_2'')$$

$$\frac{a_m}{a_C} = \frac{k_2'''}{k_2''}$$

Then

$$\frac{\gamma_m}{\gamma_C} = \frac{a_m/(m_2/m^{\circ})}{a_C/(C_2/C^{\circ})} = \frac{k_2'''(C_2/C^{\circ})}{k_2''(m_2/m^{\circ})}$$

In order to obtain a value of the ratio $k_2 : : /k_1''$, we consider dilute solutions, in which Henry's law

is followed

$$f_2 = k_2'''(C_2/C^{\circ}) = k_2''(m_2/m^{\circ}) \quad \text{and} \quad m_2 = C_2/\rho_1$$

so that

$$k_2'''(C_2/C^\circ) = k_2'' [C_2/(\rho_1 m^\circ)]$$

$$k_2'''/k_2'' = C^\circ/(\rho_1 m^\circ)$$

Then

$$\frac{\gamma_m}{\gamma_C} = \frac{C^\circ}{\rho_1 m^\circ} \frac{(C_2/C^\circ)}{(m_2/m^\circ)} = \frac{1}{\rho_1} \frac{C_2}{m_2} = \frac{1}{\rho_1} (\rho - C_2 M_2)$$

6.

$$G_{m,mix}^{E} = \Delta G_{m,mix} - X_1 RT \ln X_1 - X_2 RT \ln X_2$$

$$\left(\frac{\partial G_{m,mix}^{E}}{\partial T} \right)_{P,X} = \left(\frac{\partial \Delta G_{m,mix}}{\partial T} \right)_{P,X} - X_1 R \ln X_1 - X_2 R \ln X_2$$

$$= -\Delta S_{m,mix} - X_1 R \ln X_1 - X_2 R \ln X_2$$

$$= - S_{m,mix}^{E}$$

$$G_{m,mix}^{E} - T \left(\frac{\partial \Delta G_{m}^{E}}{\partial T} \right)_{P,X} = \Delta G_{m,mix} - X_1 RT \ln X_1 - X_2 RT \ln X_2$$

$$- T(- \Delta S_{m,mix} - X_1 R \ln X_1 + X_2 R \ln X_2)$$

$$= \Delta H_{m,mix} = H_{m,mix}^{E}$$

8.   We are interested in the transformation

S(rhombic,satd in CS_2) = S(monoclinic,satd in CS_2)

The change in the Gibbs function depends on the solubilities, the molalities in the saturated

solutions, so that the Gibbs function for this transformation will lead us to the desired molality.

$$\Delta G_m^\circ = \mu_{2\circ} + RT \ln (m_{\text{monoclinic,satd}}/m^\circ)$$

$$- [\mu_2^\circ + RT \ln (m_{\text{rhombic,satd}}/m^\circ)]$$

$$= RT \ln [(m_{\text{monoclinic,satd}})/(m_{\text{rhombic,satd}})]$$

since the dissolved solute, and hence μ_2°, is the same for both solutions.

This transformation can be expressed as the sum of the following transformations:

1. $S(\text{rhombic,s}) = S(\text{monoclinic,s})$

$$\Delta G_m^\circ = \Delta G_{m1} = 96 \text{ J mol-1}$$

2. $S(\text{monoclinic,s}) = S(\text{monoclinic,satd in } CS_2)$

$$\Delta G_{m2} = 0$$

3. $S(\text{rhombic,satd in } CS_2) = S(\text{rhombic,s})$

$$\Delta G_{m3} = 0$$

$$\Delta G_m^\circ = \Delta G_{m1} + \Delta G_{m2} + \Delta G_{m3} = 96 \text{ J mol}^{-1}$$

Therefore

$$96 \text{ J mol}^{-1} = RT \ln [(m_{\text{monoclinic,satd}})/(m_{\text{rhombic,satd}})]$$

$$= (8.3145 \text{ J mol}^{-1} \text{ K}^{-1})(298.15 \text{ K}) \ln [(m_{\text{monoclinic,satd}})/(m_{\text{rhombic,satd}})]$$

and

$$[(m_{\text{monoclinic,satd}})/(m_{\text{rhombic,satd}})] = 1.04$$

and $m_{\text{rhombic,satd}} = (22 \text{ mol kg}^{-1})/(1.04) = 21 \text{ mol kg}^{-1}$

10.

$$\frac{G_m^E}{RT} = X_1 \ln \gamma_1 + (1 - X_1) \ln \gamma_2$$

If we differentiate with respect to X_1, the result is

$$\frac{d}{dX_1}\left(\frac{G_m^{\mathrm{E}}}{RT}\right) = X_1\frac{d\ln\gamma_1}{dX_1} + \ln\gamma_1 + (1 - X_1)\left(\frac{d\ln\gamma_2}{dX_1}\right) - \ln\gamma_2$$

From the Gibbs-Duhem equation,

$$n_1 d\mu_1 = -n_2 d\mu_2 \qquad \text{or}$$

$$X_1 d\mu_1 = -X_2 d\mu_2 \qquad \text{or}$$

$$X_1\left(\frac{\partial\mu_1}{\partial X_1}\right)_{T,P} = -X_2\left(\frac{\partial\mu_2}{\partial X_1}\right)_{T,P}$$

But,

$$\mu_1 = \mu_1^{\circ} + RT\ln X_1 + RT\ln\gamma_1 \qquad \text{and}$$

$$\mu_2 = \mu_2^{\circ} + RT\ln X_2 + RT\ln\gamma_2$$

so that,

$$\left(\frac{\partial\mu_1}{\partial X_1}\right)_{T,P} = RT\left(\frac{\partial\ln X_1}{\partial X_1}\right)_{T,P} + RT\left(\frac{\partial\ln\gamma_1}{\partial X_1}\right)_{T,P} \qquad \text{and}$$

$$\left(\frac{\partial\mu_2}{\partial X_1}\right)_{T,P} = RT\left(\frac{\partial(1 - X_1)}{\partial X_1}\right)_{T,P} + RT\left(\frac{\partial\ln\gamma_2}{\partial X_1}\right)_{T,P}$$

Then,

$$X_1\left[RT\left(\frac{\partial\ln X_1}{\partial X_1}\right)_{T,P} + RT\left(\frac{\partial\ln\gamma_1}{\partial X_1}\right)_{t,P}\right] = -X_2\left[RT\left(\frac{\partial\ln X_2}{\partial X_1}\right)_{T,P} + RT\left(\frac{\partial\ln\gamma_2}{\partial X_1}\right)_{t,P}\right] \qquad \text{or}$$

$$\frac{X_1}{X_1} + X_1\left(\frac{\partial\ln\gamma_1}{\partial X_1}\right)_{T,P} = \frac{1 - X_1}{1 - X_1} - (1 - X_1)\left(\frac{\partial\ln\gamma_2}{\partial X_2}\right)_{T,P} \qquad \text{and}$$

$$X_1 \left(\frac{\partial \ln \gamma_1}{\partial X_1} \right)_{T,P} = - (1 - X_1) \left(\frac{\partial \ln \gamma_2}{\partial X_1} \right)_{T,P}$$

Therefore,

$$\frac{d}{dX_1} \left(\frac{G_m^E}{RT} \right) = \ln \gamma_1 - \ln \gamma_2 = \ln \frac{\gamma_1}{\gamma_2}$$

Chapter 17

Determination of Nonelectrolyte Activities and Excess Gibbs Functions from Experimental Data

2.   a) Table 17-1 shows values of p_1 and p_2, calculated from the relationship $p_i = P y_i$.

Table 17-1. Partial pressures of methyl *tert*-butyl ether (1) and acetonitrile (2) for solutions at 313.15 K.

x_1	p_1	p_2	x_1	p_1	p_2
0.0122	3.1727813	22.476219	0.407	37.336608	16.955392
0.0189	4.3646002	22.2164	0.4561	39.155955	16.345045
0.0263	5.7609906	22.493009	0.6203	44.550275	13.960725
0.0339	7.347672	22.232328	0.6891	46.864276	12.683724
0.0485	9.4280725	22.262927	0.7381	48.510098	11.623902
0.0669	12.689212	21.375787	0.7792	49.951006	10.580994
0.1036	17.383094	20.829906	0.8278	51.758746	9.1052544
0.1406	21.166678	20.426322	0.8539	52.746012	8.2249879
0.1652	23.780177	20.005823	0.881	53.889411	7.1335887
0.211	27.236495	19.345505	0.9119	55.216075	5.7019248
0.2372	28.969811	19.033189	0.936	56.30032	4.4466804
0.249	29.779308	18.879692	0.954	57.171699	3.3723008
0.2773	31.417254	18.522746	0.9673	57.856868	2.4861316

x_1	p_1	p_2	x_1	p_1	
0.306	32.892988	18.191012	0.9781	58.430081	1.7079192
0.3262	33.86563	17.94837	0.9846	58.806	1.194
0.3501	34.987202	17.656798	0.9902	59.091558	0.7904424
0.3742	36.021558	17.367442			

Figure 17-1 shows a plot of total pressure and partial pressures as a function of X_1, together with the Raoult's law lines for each component. We can see that Raoult's law is followed in dilute solution for each component.

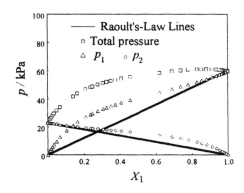

Figure 17-1. A plot of the vapor pressure data of Table 17-1, with Raoult's law lines.

b) Table 17-2 shows values of p_1/X_1 and p_2/X_2 for these solutions.

Table 17.2

x_1	p_1/x_1	p_2/x_2	x_1	p_1/x_1	p_2/x_2
0.0122	260.06404	22.753815	0.407	91.736139	28.592566
0.0189	230.93123	22.644379	0.4561	85.849497	30.051562
0.026299999	219.04907	23.100554	0.6203	71.820531	36.767776
0.033899999	216.74549	23.01245	0.6891	68.007947	40.796796
0.048499999	194.39325	23.397717	0.7381	65.722934	44.382979
0.066899998	189.67433	22.908357	0.7792	64.105501	47.921167
0.1036	167.79048	23.237289	0.8278	62.525665	52.876042

x_1	p_1/x_1	p_2/x_2	x_1	p_1/x_1	p_2/x_2
0.1406	150.54536	23.76812	0.8539	61.770713	56.296974
0.1652	143.9478	23.96481	0.881	61.168458	59.946124
0.211	129.08292	24.519017	0.9119	60.550581	64.721053
0.2372	122.13242	24.951743	0.936	60.149914	69.479381
0.249	119.59561	25.139403	0.954	59.928406	73.310887
0.2773	113.29699	25.629924	0.9673	59.812745	76.028489
0.30600001	107.49342	26.211833	0.9781	59.738351	77.987178
0.3262	103.81861	26.637533	0.9846	59.725777	77.532468
0.3501	99.934883	27.168484	0.9902	59.676386	80.657388
0.3742	96.26285	27.752384			

Figure 17-2 shows plots of p_1/X_1 and p_2/X_2 as a function of X_1. We can see that a horizontal asymptote is found for each component at $X = 1$, so that a Raoult's law standard state is feasible. There are not horizontal asymptotes at $X = 0$, so that Henry's law standard states are not feasible by that strong requirement. For component 2, we can extrapolate the curve to a finite value at $X = 0$ to find a value for the Henry's law constant; even that weak requirement is not found for component 1.

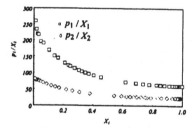

Figure 17-2. A plot of p_i/X_i for the data of Table 17-2.

c) The Raoult's law standard states are permitted by the data. The activities are given by $a_i = p_i/p^*$ and the activity coefficients by a_i/X_i. Table 17-3 shows the calculated values, and Figure 17-3 shows a plot of the activities as a function of X_1. Figure 17-4 shows a corresponding plot of the activity coefficients.

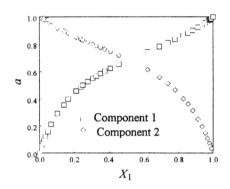

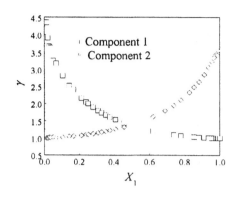

Figure 17-3. A plot of a_i for the calculations

Table 17-3.

Figure 17-4. A plot of γ_i for the calculations of

of Table 17-3.

Table 17-3.

X_1	a_1	a_2	γ_1	γ_2
0.0122	0.053077846	0.98988015	4.3506431	1.0021058
0.0189	0.073015929	0.97843741	3.8632767	0.99728612
0.0263	0.096376315	0.99061963	3.6644987	1.0173766
0.0339	0.1229201	0.97913891	3.6259617	1.0134964
0.0485	0.15772338	0.98048655	3.2520284	1.0304641
0.0669	0.21227938	0.94141582	3.173085	1.008912
0.1036	0.2908039	0.91737454	2.8069874	1.0233986
0.1406	0.35409993	0.89960021	2.5184917	1.0467771
0.1652	0.39782148	0.88108092	2.4081203	1.0554395
0.211	0.45564266	0.85199967	2.1594439	1.0798475

X_l	a_l	a_2	γ_1	γ_2
0.2372	0.4846395	0.83824494	2.0431682	1.0989053
0.249	0.49818168	0.83148472	2.0007296	1.1071701
0.2773	0.52558308	0.81576438	1.8953591	1.1287732
0.306	0.5502708	0.80115443	1.7982706	1.1544012
0.3262	0.56654226	0.79046814	1.7367942	1.1731495
0.3501	0.58530518	0.77762695	1.6718228	1.1965332
0.3742	0.60260905	0.76488337	1.610393	1.2222489
0.407	0.62460868	0.74673618	1.5346651	1.2592516
0.4561	0.65504476	0.71985574	1.4361867	1.3235075
0.6203	0.74528699	0.61484738	1.2014944	1.6192978
0.6891	0.78399819	0.55860671	1.1377132	1.7967408
0.7381	0.81153135	0.51193086	1.099487	1.9546807
0.7792	0.83563648	0.46599989	1.0724288	2.1105068
0.8278	0.86587837	0.40100654	1.0459995	2.3287255
0.8539	0.88239447	0.36223852	1.0333698	2.4793876
0.881	0.90152254	0.31417197	1.0232946	2.6401006
0.9119	0.92371646	0.25111974	1.0129581	2.8503943
0.936	0.94185492	0.19583724	1.0062553	3.0599569
0.954	0.95643233	0.14852025	1.0025496	3.2287011
0.9673	0.96789461	0.10949228	1.0006147	3.3483876
0.9781	0.97748395	0.07521885	0.99937016	3.4346507

X_1	a_1	a_2	γ_1	γ_2
0.9846	0.98377275	0.05258522	0.99915981	3.4146247
0.9902	0.98854988	0.03481205	0.99833355	3.55225

The activity coefficients are all greater than or equal to 1 because the activities are all greater than or equal to the corresponding mole fractions, consistent with the positive deviations from Raoult's law.

d)   The values of G_m^E are calculated from Equation 16-63

$$G_m^E = RT \left[X_1 \ln \gamma_1 + (1 - X_1) \ln \gamma_2 \right]$$

and are shown in Table 17-4 and plotted in Figure 17-5.

Table 17-4.

X_1	G_m^E	X_1	G_m^E
0.0122	52.115144	0.407	809.79819
0.0189	59.565652	0.4561	826.80445
0.0263	132.60545	0.6203	772.97874
0.0339	147.41796	0.6891	705.82619
0.0485	223.26296	0.7381	639.29997
0.0669	222.68984	0.7792	571.2695
0.1036	332.38573	0.8278	475.93511
0.1406	440.42628	0.8539	418.38588
0.1652	495.2965	0.881	353.61796
0.211	580.75038	0.9119	270.83942
0.2372	628.58926	0.936	201.5626
0.249	648.68675	0.954	146.70445

X_I	G_m^E	X_I	G_m^E
0.2773	689.58501	0.9673	104.43824
0.306	726.98633	0.9781	68.754233
0.3262	749.01855	0.9846	47.08673
0.3501	772.07615	0.9902	28.043775
0.3742	791.23804		

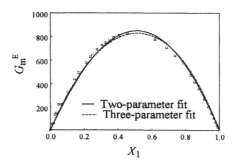

Figure 17-5. A plot of G_m^E from the calculations of Table 17-4, with a two-parameter fit and a three-parameter fit of the Redlich-Kister equation.

We can see from Figure 17-5 that both the two-parameter fit of the Redlich-Kister equation and the three-parameter fit are close.

e) The values of $\ln \gamma_1/\gamma_2$ are shown in Table 17-5 and plotted in Figure 17-6.

Table 17-5.

X_I	$\ln\gamma_1/\gamma_2$	X_I	$\ln\gamma_1/\gamma_2$
0.0122	1.46822	0.407	0.19779461
0.0189	1.3542333	0.4561	0.081706055
0.0263	1.2814641	0.6203	-0.29842647
0.0485	1.1492697	0.7381	-0.57538314
0.0669	1.1458317	0.7792	-0.67700216

X_1	$\ln\gamma_1/\gamma_2$	X_1	$\ln\gamma_1/\gamma_2$
0.1036	1.0089827	0.8278	-0.80034826
0.1406	0.87794422	0.8539	-0.87518646
0.1652	0.82488919	0.881	-0.94778959
0.211	0.6930309	0.9119	-1.0345825
0.2372	0.62018717	0.936	-1.1121651
0.249	0.59170466	0.954	-1.1695335
0.2773	0.51827695	0.9673	-1.2078644
0.306	0.44324367	0.9781	-1.2345453
0.3262	0.39234897	0.9846	-1.2289081
0.3501	0.33448615	0.9902	-1.269249
0.3742	0.2757857		

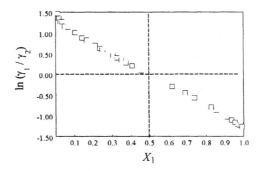

Figure 17-6. A plot of $\ln \gamma_1/\gamma_2$ from the calculations of Table 17-5.

f) The curve is neither horizontal nor linear. Therefore, we require a Redlich-Kister polynomial that requires more than one parameter. We can see from Figure 17-5 that the three-parameter equation and the two-parameter equation provide a close fit. The three-parameter function is

$$G_m^E = RT\, X_1\, X_2\, [1.270 - 0.0749\, (X_1 - X_2) + 0.153\, (X_1 - X_2)^2\,]$$

The positive sign of B is consistent with the positive deviation from the Raoult's law lines.

4.   Table 17-6 shows the data and calculations for this exercise.

Table 17-6.

X_{In}	$\mathscr{E}$	$\mathscr{E}'$	$\ln \gamma$	γ	a
0.7000	0.2668	31.5095	-3.5195	0.0296	0.0207
0.6441	0.2643	31.3008	-3.3108	0.0365	0.0235
0.6360	0.26303	31.1652	-3.1752	0.0418	0.0266
0.5222	0.25895	30.8859	-2.8959	0.0552	0.0289
0.5000	0.25831	30.8546	-2.8646	0.0570	0.0285
0.3998	0.2518	30.3181	2.3281	0.0975	0.0390
0.3040	0.2455	29.8565	-1.8665	0.1547	0.0470
0.3000	0.24535	29.8522	-1.8622	0.1533	0.0466
0.2000	0.23706	29.2897	-1.2997	0.2726	0.0545
0.1626	0.2347	29.2211	-1.2311	0.2920	0.0475
0.1000	0.2265	28.7451	-0.7551	0.4700	0.0470
0.0841	0.22331	28.5505	-0.5605	0.5709	0.0480
0.0500	0.2179	28.4388	-0.4488	0.6384	0.0319
0.0388	0.2144	28.2837	-0.2937	0.7445	0.0289
0.0100	0.2014	28.1262	-0.1362	0.8726	0.0087
0.0092	0.2007	28.1209	-0.1309	0.8773	0.0081
0.0009	0.1790	27.8705	0.1195	1.1269	0.0011

The quantity $\mathscr{E}'$ is

$$\frac{n \mathscr{F} \mathscr{E}}{RT} + \ln X_{\mathrm{In}}$$

A plot of $\mathscr{E}'$ as a function of X_{In} is shown in Figure 17-7. A straight line was fitted by least squares to

the most dilute 8 points, as shown in the figure. The intercept at $X_{\mathrm{In}} = 0$ is equal to $\ln a_2'$ of Equation

17-24, and it has a value of 27.99. Then $\ln \gamma_2 = 27.99 - \mathscr{E}'$. The activity is equal to $X \gamma$.

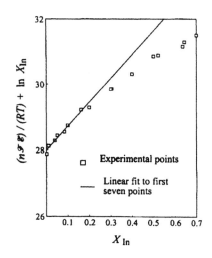

Figure 17-7. A plot of $\mathscr{E}'$, with an extrapolation to $X_{\mathrm{In}} = 0$.

6.   Table 17-7 shows calculated values of p_1 and p_2 as well as values of a_1 and γ_1 and a_2 and γ_2

calculated on the basis of Raoult's law. Figure 17-8 shows that the data do not follow Raoult's law, but

that there is a region in dilute solution for both components that follows Raoult's law and provides a

basis for using Raoult's law to calculate activities.

Table 17-8 shows calculated values of p_i/X_i for both components and Figure 17-9 shows that a

plot of p_i/X_i has a horizontal asymptote at X_i near 1 for both components, supporting the choice of

Raoult's law to calculate the activities. Although there are no horizontal asymptotes near $X_i = 0$, the

strong basis for Henry's law, we can see that the curves have a finite limit and provide a weak basis for

a Henry's law calculation of activities.

Table 17-8 also shows calculated values of G_m^E as a function of X_1, and values of $\ln (\gamma_1/\gamma_2)$ that provide a basis of deciding which Redlich-Kister equation to fit to the curve of G_m^E. The plot of $\ln(\gamma_1/\gamma_2)$ in Figure 17-11 shows that the solution is not regular, since the points do not fall on a straight line, so we tried a two-parameter fit of the Redlich-Kister equation. The resulting equation is

$$G_m^E = RT X_1 X_2 \left[0.3615 + 0.0233 (2X_1 - 1) \right]$$

and the fitted curve in Figure 17-10 shows an excellent fit. Calculated values of G_m^E are also shown in Table 17-8 for comparison to the experimental values.

Table 17-7.

p_2	p_1	a_1	a_2	γ_1	γ_2
30.735	0.000	0.000	1.000		1.000
30.170	1.783	0.027	0.982	1.370	1.002
29.527	3.676	0.056	0.961	1.348	1.003
28.887	5.621	0.086	0.940	1.326	1.005
28.488	6.696	0.102	0.927	1.321	1.005
27.567	9.121	0.139	0.897	1.291	1.006
26.061	13.384	0.205	0.848	1.255	1.013
24.239	18.196	0.278	0.789	1.216	1.023
22.534	22.553	0.345	0.733	1.180	1.036
21.528	25.040	0.383	0.700	1.166	1.043
20.046	28.657	0.438	0.652	1.140	1.060
18.209	32.825	0.502	0.592	1.111	1.081
16.302	36.972	0.565	0.530	1.084	1.109
14.199	41.310	0.632	0.462	1.060	1.143

p_2	p_1	a_1	a_2	γ_1	γ_2
13.376	42.945	0.657	0.435	1.052	1.158
10.450	48.527	0.742	0.340	1.029	1.218
8 624	51.767	0.792	0.281	1.019	1.259
6.243	55.816	0.853	0.203	1.009	1.321
4.042	59.410	0.908	0.132	1.004	1.381
2.770	61.353	0.938	0.090	1.002	1.422
1.176	63.774	0.975	0.038	1.001	1.477
0.000	65.403	1.000	0.000	1.000	

Table 17-8.

p_1/x_1	p_2/x_2	G_m^E	$\ln(\gamma_1/\gamma_2)$	$G_m^E(\text{calc})$
	30.735			
89.597	30.783	19.934	0.3132	17.165
88.143	30.812	38.048	0.2959	35.275
86.749	30.888	58.808	0.2775	53.664
86.394	30.882	66.534	0.2736	63.419
84.450	30.905	83.340	0.2501	85.811
82.109	31.137	122.836	0.2145	122.432
79.529	31.430	158.820	0.1732	159.749
77.155	31.842	187.910	0.1299	188.868
76.271	32.051	201.431	0.1118	202.298
74.530	32.569	220.088	0.0727	218.712

p_1/x_1	p_2/x_2	G_m^E	$\ln(\gamma_1/\gamma_2)$	$G_m^E(\text{calc})$
72.638	33.222	230.742	0.0271	230.920
70.869	34.083	233.983	-0.0231	234.743
69.323	35.138	227.488	-0.0757	228.694
68.811	35.585	222.319	-0.0957	223.601
67.315	37.445	194.420	-0.1687	194.125
66.616	38.689	168.026	-0.2118	168.298
65.961	40.593	128.025	-0.2697	127.538
65.661	42.457	87.933	-0.3191	85.021
65.506	43.693	60.917	-0.3502	58.839
65.472	45.391	28.510	-0.3889	25.114
65.403				

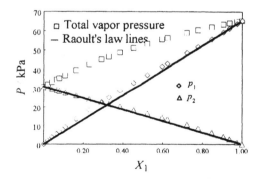

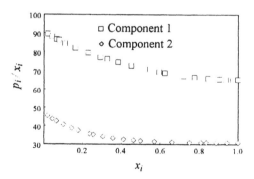

Figure 17-8. A plot of vapor pressures and Raoult's law lines for the data of Table 17-7.

Figure 17-9. A plot of p_i/X_i for the data of Table 17-8.

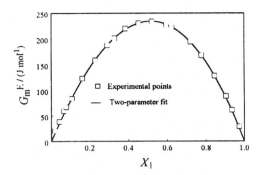

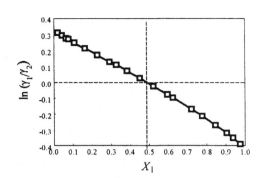

Figure 17-10. A plot of G_m^E for the data of 17-8.

Figure 17-11. A plot of $\ln \gamma_1/\gamma_2$ for the data of Table Table 17-8, together with a two-parameter Redlich-Kister fit.

Chapter 18

Calculation of Partial Molar Quantities and Excess Molar Quantities from Experimental Data: Volume and Enthalpy

2.   a) V is the volume of solution that contains 1 kg of solvent. That is,

$$V = \frac{1.000 + m_2 M_2}{\rho}$$

where 1.000 has the units of [kg (kg solvent)$^{-1}$], m_2 has the units [moles solute (kg solvent)$^{-1}$], M_2 has the units [kg solute mol^{-1}], ρ has the units [kg (unit volume)$^{-1}$], and V has the units [volume (kg solvent)$^{-1}$]. ω_1 is the mass of solvent. Thus,

$$V\rho = 1.000 + m_2 M_2$$

$$V\left(\frac{\partial \rho}{\partial m_2}\right)_{T,P} + \rho\left(\frac{\partial V}{\partial m_2}\right)_{T,P,\omega_1 = 1\text{ kg}} = M_2$$

where ρ is a function only of m_2 at constant T and P, whereas V is a function of m_2 and ω_1.

Therefore

$$V_{m2} = \left(\frac{\partial V}{\partial n_2}\right)_{T,P,n_1} = \left(\frac{\partial V}{\partial m_2}\right)_{T,P,\omega_1 = 1\text{kg}}$$

$$= \frac{M_2 - V\left(\frac{\partial \rho}{\partial m_2}\right)_{T,P}}{\rho}$$

b)

$$V = n_1 V_{m1} + n_2 V_{m2}$$

where n_1 has units of (moles solvent)/(kg of solvent), V_{m1} has units of volume per mole of solvent, n_2 has units of (moles solute)/(kg of solvent), equal to m_2, and V_{m2} has units of volume per mole of solute. Then

$$V_{m1} = \frac{V - n_2 V_{m2}}{n_1} = \frac{V - m_2 V_{m2}}{(1.000 \text{ kg solvent})/M_1} = \frac{M_1 (V - m_2 V_{m2})}{1.000 \text{ kg solvent}}$$

$$= \frac{M_1}{1.000 \text{ kg solvent}} \left[V - m_2 \frac{M_2 - V\left(\frac{\partial \rho}{\partial m_2}\right)_{T,P}}{\rho} \right]$$

$$= \frac{M_1}{\rho(1.000 \text{ kg solvent})} \left[V\rho - m_2 M_2 + m_2 V\left(\frac{\partial \rho}{\partial m_2}\right)_{T,P} \right]$$

But $V\rho$ is the mass of the solution per kilogram of solvent, and $m_2 M_2$ is the mass of the solute per kilogram of solvent. Therefore, their difference is the mass of solvent, or 1.000 kg. Thus,

$$V_{m1} = \frac{M_1}{\rho} \left[1.000 + m_2 V \left(\frac{\partial \rho}{\partial m_2}\right)_{T,P} \right]$$

4.

Table 18-1.

WT% EtOH	n_2/n_1	n_1/n_2	ρ/(g cm^{-3})	V/ [cm^3 /(mol H2O)]	$\Delta V/\Delta n_2$	V/ [cm^3 /(mol EtOH)]	$\Delta V/\Delta n_1$
20	0.097761	10.2290	0.96639	23.3024		238.3603	
25	0.130349	7.67174	0.95895	25.0488	53.59	192.1677	18.063
30	0.167591	5.96691	0.95067	27.0717	54.32	161.5345	17.968
35	0.210563	4.74917	0.94146	29.4394	55.10	139.8126	17.838
40	0.260697	3.83587	0.93148	32.2344	55.75	123.6468	17.700
45	0.319947	3.12552	0.92085	35.5707	56.31	111.1770	17.555
50	0.391046	2.55725	0.90985	39.6008	56.68	101.2690	17.435
55	0.477945	2.09229	0.89850	44.5567	57.03	93.2257	17.299
60	0.586569	1.70483	0.88699	50.7768	57.26	86.5658	17.188
65	0.726228	1.37698	0.87527	58.8076	57.50	80.9769	17.047
70	0.912440	1.09596	0.86340	69.5522	57.70	76.2266	16.904

Table 18-1 is Table 18-1 in text.

a) For 20 wt% ethanol:

$$\frac{n_2}{n_1} = \frac{\text{moles C}_2\text{H}_5\text{OH}}{\text{moles H}_2\text{O}} = \frac{(20 \text{ g H}_2\text{O})/(46.0698 \text{ g mol}^{-1})}{(80 \text{ g H}_2\text{O})/(18.0154 \text{ g mol}^{-1})}$$

$$= \frac{0.4341239}{4.440645} = 0.097761$$

$$\frac{n_1}{n_1} = \frac{\text{moles H}_2\text{O}}{\text{moles C}_2\text{H}_5\text{OH}} = 10.228982$$

The volume per mole of H_2O is:

$$V = \frac{\dfrac{100 \text{ g}}{0.96639 \text{ g cm}^{-3}}}{4.440645 \text{ mol H}_2\text{O}} = 23.3024 \text{ cm}^3 \text{ (mol H}_2\text{O)}^{-1}$$

The volume per mole of C_2H_5OH is:

$$V = \frac{\dfrac{100 \text{ g}}{0.96639 \text{ g cm}^{-3}}}{0.4341239 \text{ mol EtOH}} = 238.3603 \text{ cm}^3 \text{ (mol EtOH)}^{-1}$$

For 25 wt% ethanol:

$$\frac{n_2}{n_1} = \frac{\text{moles C}_2\text{H}_5\text{OH}}{\text{moles H}_2\text{O}} = \frac{(25 \text{ g H}_2\text{O})/(46.0698 \text{ g mol}^{-1})}{(75 \text{ g H}_2\text{O})/(18.0154 \text{ g mol}^{-1})}$$

$$= \frac{0.5426548}{4.163105} = 0.130349$$

$$\frac{n_1}{n_2} = \frac{\text{moles H}_2\text{O}}{\text{moles C}_2\text{H}_5\text{OH}} = 7.671737$$

The volume per mole of H_2O is:

$$V = \frac{\dfrac{100 \text{ g}}{0.95895 \text{ g cm}^{-3}}}{4.163105 \text{ mol H}_2\text{O}} = 25.0488 \text{ cm}^3 \text{ (mol H}_2\text{O)}^{-1}$$

The volume per mole of EtOH is:

$$V = \cfrac{\cfrac{100 \text{ g}}{0.95895 \text{ g cm}^{-3}}}{0.5426548 \text{ mol H}_2\text{O}} = 192.1677 \text{ cm}^3 \text{ (mol C}_2\text{H}_5\text{OH)}^{-1}$$

$$\text{and} \quad \frac{\Delta V}{\Delta n_2} = \frac{25.0488 - 23.3024}{0.130349 - 0.097761} = 53.59$$

$$\frac{\Delta V}{\Delta n_1} = \frac{192.1677 - 238.3603}{7.67174 - 10.22898} = 18.063$$

b) Figure 18-1 shows the volume per mole of H_2O plotted against n_2/n_1.

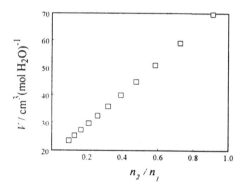

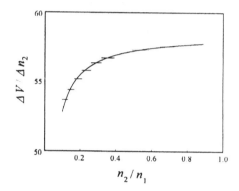

Figure 18-1. A plot of volume per mole of H_2O against n_2/n_1.

Figure 18-2. A chord plot of $\Delta V/\Delta n_2$ against n_2/n_1.

c) Figure 18-2 shows a chord plot of $\Delta V/\Delta n_2$ against n_2/n_1. The graphical differentiation was done on a scale large enough to use all the precision of the data.

d) It is easier said than done to fit the volume per mole of H_2O as a function of n_2/n_1 to a polynomial, with useful results. If we fit the data to a linear equation, since the Figure looks fairly linear, we obtain the equation

$$V/[cm^3 \ (molH_2O)^{-1}] = 17.5156 + 56.8285 \ (n_2/n_1)$$

However, the residuals are not randomly distributed around 0, indicating that the curve is not

linear, even though the correlation coefficient is 0.99996, and the sum of squares of deviations

is 0.17. We can get successively better fits with quadratic and cubic polynomials, as indicated by

the residuals, the correlation coefficients, and the sum of squares, but the derivatives of these

functions do not fit the graph of $\Delta V/\Delta n_2$ as shown in Figure 18-2. The curve that we found to

fit the chord plot in Figure 18-2 is

$$\frac{\Delta V}{\Delta (n_2/n_1)} = \frac{58.398 \ (n_2/n_1)}{0.01076 + (n_2/n_1)}$$

This result implies that V as a function of n_2/n_1 should be the integral of this equation within an

additive constant of integration. The best nonlinear fit of such an equation to the data in the

problem is

$$V/[cm^3 \ (mol \ H_2O)^{-1}] = 16.129 + 58.492 \ [n_2/n_1 - 0.01128 \ \log \ (n_2/n_1 + 0.01128)]$$

The goodness of fit is shown in Figure 18-3. The goodness of fit is also indicated by the sum of

squares of the deviations equal to 0.000115 and the correlation coefficient equal to 0.9999999.

The results of this calculation indicate the superiority of graphical differentiation to the arbitrary

fitting of data to a function that may look like a good fit, but does not yield an appropriate

derivative. The graphical differentiation follows the data without any prior assumption of a

functional form. I am not aware of a model that would lead to the best fit we obtained after

ascertaining the form of the chord plot.

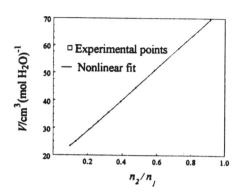

Figure 18-3. A nonlinear fit of volume per mole of H_2O against n_2/n_1.

e) From a large scale graph like the one in Figure 18-2, $V_{m2} = 56.82$ cm^3(mol EtOH)$^{-1}$.

From the best fit equation for V as a function of n_2/n_1, as in Figure 18-3,

$$V_{m2} = \frac{(58.491)(0.319947)}{0.0113 + 0.319947} = 56.50 \text{ cm}^3(\text{mol H}_2\text{O})^{-1}$$

which is good agreement.

f) The volume per mole of ethanol is plotted against n_1/n_2 in Figure 18-4.

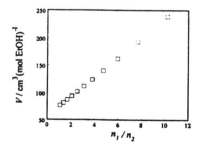

Figure 18-4. A plot of volume per mole of ethanol against n_1/n_2.

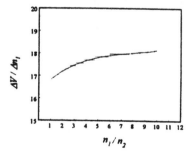

Figure 18-5. A chord plot of $\Delta V/\Delta n_1$ against n_1/n_2.

g) Figure 18-5 shows a chord plot of $\Delta V/\Delta n_1$ against n_1/n_2.

h) Figure 18-6 shows a plot of $V/[cm^3(mol\ EtOH)^{-1}]$ against n_1/n_2 that is fitted to a quartic polynomial

$$V/[cm^3\ (mol\ H_2O)^{-1}] = 57.7804 + 16.5896\ (n_1/n_2)$$

$$+ 0.188485\ (n_1/n_2)^2 - 0.010471\ (n_1/n_2)^3$$

$$+ 0.0002162\ (n_1/n_2)^4$$

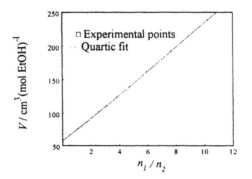

Figure 18-6. A plot of $V/[cm^3(mol\ EtOH)^{-1}]$ against n_1/n_2.

i) From a large scale graph like the one in g), $V_{m1} = 17.502\ cm^3\ (mol\ H_2O)^{-1}$.

From the equation in h),

$$V_{m1} = \left(\frac{\partial V}{\partial n_1}\right)_{T,P,n_2}$$

$$= 16.58955 + 0.37697(n_1/n_2) - 0.031410(n_1/n_2)^2$$

$$+ 0.0008648(n_1/n_2)^3 = 17.487\ cm^3\ (mol\ H_2O)^{-1}$$

which shows good agreement between the two results.

However, when we plot the derivative curve of the quartic polynomial in h), as shown in Figure 18-7, the agreement is not good at low values of n_1/n_2. Again we see the advantage of graphical differentiation over fitting an arbitrary function to obtain a derivative, even when the function fits the volume very well.

j)

$$n_1 = \frac{(500 \text{ g})(0.55)}{18.0154 \text{ g mol}^{-1}} = 15.2647 \text{ mol}$$

$$n_2 = \frac{(500)(0.45)}{46.0698 \text{ g mol}^{-1}} = 4.88389 \text{ mol}$$

$$V = n_1 V_{m1} + n_2 V_{m2}$$

$$= (15.2647 \text{ mol})(17.502 \text{ cm}^3 \text{ mol}^{-1}) + (4.88389 \text{ mol})(56.82 \text{ cm}^3 \text{ mol}^{-1})$$

$$= 542.92 \text{ cm}^3$$

From the measured density,

$$V = \frac{500\text{g}}{0.92058 \text{ g cm}^{-3}} = 542.98 \text{ cm}^{-3}$$

which is excellent agreement.

k) The basis for this calculation is a rearranged form of Eq.18-37,

$$dV_{m2} = -\frac{n_1}{n_2} dV_{m1}$$

which follows from the Gibbs-Duhem equation. The value required is obtained by integration as follows:

$$V_{m2}(65\%) - V_{m2}(25\%) = \int_{25\%}^{65\%} -\frac{n_1}{n_2} dV_{m1}$$

The numerical integration of the integral above is carried out in Table 18-2, where the values of V_{m1} as a function of n_1/n_2 are read from the chord plot in Figure 18-5.

Table 18-2.

$V_{ml}/$ cm^3(mol)$^{-1}$	n_1/n_2	$(n_1/n_2)_{avg}$	ΔV_{ml}	$\Sigma(n_1/n_2)_{avg}\Delta V_{ml}$
17.032	1.500	1.440	0.072	0.104
17.209	2.000	1.750	0.177	0.413
17.349	2.500	2.250	0.140	0.728
17.470	3.000	2.750	0.121	1.061
17.562	3.500	3.250	0.092	1.360
17.650	4.000	3.750	0.088	1.690
17.732	4.500	4.250	0.082	2.039
17.802	5.000	4.750	0.070	2.371
17.858	5.500	5.250	0.056	2.665
17.907	6.000	5.750	0.049	2.947
17.949	6.500	6.250	0.042	3.209
17.978	7.000	6.750	0.029	3.405
18.000	7.500	7.250	0.022	3.565
18.007	7.670	7.585	0.007	3.618

The initial value of n_1/n_2 corresponds to 65% ethanol and the final value corresponds to 25%

ethanol. Therefore

$$V_{m1}(65\%) - V_{m1}(25\%) = - (0.000 - 3.618) = 3.618 \text{ cm}^3 \text{ mol}^{-1}$$

1) The value of V_{m2} for 25% ethanol from the chord-area graph in Figure 18-2 is 54.14 cm^3 mol^{-1}. Therefore, the value of V_{m2} for 65% ethanol is 57.76 cm^3 mol^{-1}. The corresponding value from the chord-area graph is 57.52 cm^3 mol^{-1}.

6.   Table 18-3, taken from a spreadsheet, shows the calculation of $V_m = V/(n_1+n_2)$ and of X_1.

Table 18-3.

wt%EtOH	$\rho/(g \text{ cm}^{-3})$	$V/(\text{cm}^3 \text{ g}^{-1})$	n_1+n_2	$V/(n_1+n_2)$	X_1
20	0.96639	1.03478	0.0487477	21.2272	0.910945
25	0.95895	1.04281	0.0470576	22.1602	0.884683
30	0.95067	1.05189	0.0453675	23.1860	0.856464
35	0.94146	1.06218	0.0436774	24.3187	0.826064
40	0.93148	1.07356	0.0419873	25.5687	0.793212
45	0.92085	1.08595	0.0402972	26.9486	0.757606
50	0.90985	1.09908	0.0386071	28.4684	0.718884
55	0.89850	1.11297	0.0369170	30.1478	0.676615
60	0.88699	1.12741	0.0352269	32.0042	0.630291
65	0.87527	1.14250	0.0335368	34.0671	0.579298
70	0.86340	1.15821	0.0318468	36.3683	0.522892

Figure 18-7 shows a plot of V_m against X_1 from the calculations above, along with the tangent line at 45% ethanol. The curve was fitted to a quartic equation,

$$V_m = 62.8100 - 71.6746\ X_1 + 70.9018\ X_1^2 - 74.4773\ X_1^3 + 30.7472\ X_1^4$$

The slope at 45% ethanol, $X_1 = 0.757606$, was calculated to be -39.0048, and the equation of the

tangent line is

$$y = 56.50 - 39.00\ X_1$$

Thus, the intercept at $X_1 = 0$ is 56.50 cm^3 mol^{-1} = V_{m2}, and the intercept at $X_1 = 1$ is 17.50 cm^3

mol^{-1} = V_{m1} . The corresponding values read from a large scale graph with an eye-fit tangent line

are 55.6 cm^3 mol^{-1} and 17.4 cm^3 mol^{-1}.

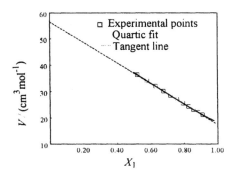

Figure 18-7. A tangent plot to determine V_{m1} and V_{m2}.

8.   a) The data are given as specific heat, c_P, or heat capacity per gram of solution. Therefore the

total heat capacity of the solution, C_P, is given by

$$C_P = c_P\,(1000 + 18.016\ m_2)$$

The data and calculated values are shown in Table 18-4.

Table 18-4.

m_2	c_P	C_P	C_{Pm2}	C_{Pm1}
0.0000	1.0000	1000.0	35.656	18.020
0.2014	0.9922	1007.2	35.942	18.021
0.4107	0.9844	1014.8	36.232	18.022
0.7905	0.9711	1028.7	36.745	18.028
1.289	0.9547	1047.0	37.386	18.037
1.7632	0.9405	1065.0	37.959	18.055
2.6537	0.9167	1099.3	38.929	18.094
4.3696	0.8790	1167.3	40.356	18.182
6.1124	0.8489	1238.4	41.099	18.251

A quartic polynomial for C_P as a function of m_2 obtained by the method of linear least squares

is

$$C_P = 1000.016 + 35.6562\,m_2 + 0.714856\,m_2^2 - 0.020999\,m_2^3 - 0.001032\,m_2^4$$

The experimental points and the fitted curve are shown in Figure 18-8. Data from F. T. Gucker,

Jr., W. L. Ford, and C. E. Moser, *J. Phys. Chem.* **45**, 309 (1941).

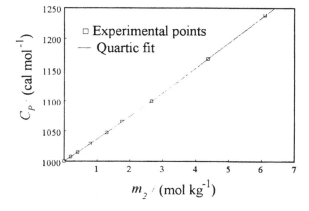

Figure 18-8. Experimental values of C_P as a function of m_2 for aqueous solutions of glycolamide. The curve is that of the quartic polynomial.

b)

$$C_{Pm2} = \left(\frac{\partial C_P}{\partial n_2}\right)_{T,P,n_1} = \left(\frac{\partial C_P}{\partial m_2}\right)_{T,P,1\ \text{kg solvent}}$$

$$= 35.6562 + 1.42971\,m_2 - 0.062997\,m_2^2 - 0.004128\,m_2^3$$

c) The calculated values of C_{Pm2} are plotted in Figure 18-9.

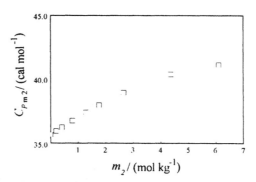

Figure 18-9. A plot of C_{Pm2} as a function of m_2.

d)

$$C_{Pm1} - C_{Pm1}^{\circ} = - \int_{C_{Pm2}^{\circ}}^{C_{Pm2}} \frac{n_2}{n_1}\,dC_{Pm2}$$

$$= - \int_{0}^{m_2} \frac{m_2}{55.51}\,(1.4297 - 0.1260\,m_2 - 0.01238\,m_2^3)\,dm_2$$

$$= 0.01288\,m_2^2 - 0.0007566\,m_2^3 - 0.00005577\,m_2^3$$

Since $C_{Pm1}{}^{\circ}$ is equal to 18.016 cal mol⁻1, the result is

$$C_{Pm1} = 18.016 + 0.01288\,m_2^2 - 0.0007566\,m_2^3 - 0.00005577\,m_2^4$$

The calculated values of C_{Pm1} at the experimental values of m_2 are shown in Figure 18-11, and are given in Table 18-2.

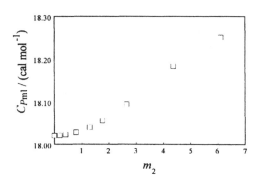

Figure 18-10. A plot of C_{Pm1} as a function of m_2.

10)

$$V_{m2} = 62.3 \text{ cm}^3 \text{ mol}^{-1}$$

Therefor, $V_{m1} = V_{m1}^\bullet$

$$V_{m2}^\bullet = \frac{253.809 \text{ g mol}^{-1}}{4.93 \text{ g cm}^{-3}} = 51.48 \text{ cm}^3 \text{ mol}^{-1}$$

$$V_{m1}^\bullet = \frac{32.04 \text{ g mol}^{-1}}{0.7865 \text{ g cn}^{-3}} = 40.74 \text{ cm}^3 \text{ mol}^{-1}$$

a)

$$\Delta V_m = V_{m2} - V_{m2}^\bullet = (62.3 - 51.48) \text{ cm}^3 \text{ mol}^{-1} = 10.8 \text{ cm}^3 \text{ mol}^{-1}$$

b)

$$\Delta V_m = 10.8 \text{ cm}^3 \text{ mol}^{-1}$$

c) $$\Delta V_m = 0$$

d) $$\Delta V_m = 0$$

e) $$\Delta V^m = V^{m2} - V_{m2}^\bullet = 10.8 \text{ cm}^3 \text{ mol}^{-1}$$

12.

$$L_{m1} = H_{m1} - H_{m1}^{\bullet}$$

$$\left(\frac{\partial L_{m1}}{\partial T}\right)_P = \left(\frac{\partial H_{m1}}{\partial T}\right)_P - \left(\frac{\partial H_{m1}^{\bullet}}{\partial T}\right)_P = C_{Pm1} - C_{Pm1}^{\bullet}$$

14.   a) For a solution containing 1 kg of solvent, we can write Equation 18-19 as,

$$L_{m2} = \left(\frac{\partial \Delta H}{\partial m_2}\right)_{n_1} - \lim_{n_2 \to 0}\left(\frac{\partial \Delta H}{\partial m_2}\right)_{n_1}$$

$$= 923 \text{ cal mol}^{-1} + 3/2(476.1 \text{ cal mol}^{-1.5} \text{ kg}^{0.5})m_2^{0.5}$$

$$- 2(726.1 \text{ cal mol}^{-2} \text{ kg})m_2 + 5/2(243.5 \text{ cal mol}^{-2.5} \text{ kg}^{1.5})m_2^{1.5}$$

$$- 923 \text{ cal mol}^{-1}$$

$$= (714.2 \text{ cal mol}^{-1.5} \text{ kg}^{0.5})m_2^{0.5} - (1452.2 \text{ cal mol}^{-2} \text{ kg})m_2$$

$$+ (608.8 \text{ cal mol}^{-2.5} \text{ kg}^{1.5})m_2^{1.5}$$

The calculated values of L_{m2} are shown in Table 18-5.

b) Similarly, for a solution containing 1 kg of solvent, we can write Equation 18-38 for relative

partial molar enthalpies as,

$$L_{m1} = \int_{L_{m2}^{\circ}}^{L_{m2}} \frac{m_2}{n_1/kg} dL_{m2}$$

$$= \int_0^{m_2} \frac{m_2}{55.51 \text{ mol kg}^{-1}} dL_{m2}$$

$$= \int_0^{m_2} \frac{1}{55.51 \text{ mol kg}^{-1}}[(357.1 \text{ cal mol}^{-1.5} \text{ kg}^{0.5})m_2^{0.5} - (1452.2 \text{ cal mol}^{-2} \text{ kg})m_2 + (913.2 \text{ cal}$$

$$= \int_0^{m_2} [(6.43 \text{ cal mol}^{-2.5} \text{ kg}^{1.5}) m_2^{0.5} - (26.16 \text{ cal mol}^{-3.0} \text{ kg}^{2.0}) m_2 + (16.45 \text{ cal mol}^{-3.5} \text{ kg}^{2.5}) m_2^{1.5}]$$

$$= 4.289 \text{ cal mol}^{-3.0} \text{ kg}^{1.5}) m_2^{1.5} - 13.08 \text{ cal mol}^{-3.5} \text{ kg}^{2.5} m_2^2 + 6.58 \text{ mol}^{-4.5} \text{ kg}^{3.0} m_2^{2.5}$$

The values calculated for L_{m1} are also shown in Table 18-5, taken from a spreadsheet.

Table 18-5.

$m_2/(\text{mol/kg})$	$L_{m2}/(\text{cal/mol})$	$L_{m1}/(\text{cal/mol})$
0.01	57.51	-0.003047
0.10	99.88	-0.025638

c) The units of the ΔH we calculate are calories per kg of solvent as a function of the molality of solute. If we divide that quantity by the molality, the number of moles of solute per kg of solvent, we obtain calories per mole of solute at molality m; that is, in enough solvent to have a molality of m with one mole of solute. The equation is

$$\frac{\Delta H}{m_2} = 923 \text{ cal mol}^{-1} + (476.1 \text{ cal mol}^{-1.5} \text{ kg}^{0.5}) m_2^{0.5}$$

$$- (726.1 \text{ cal mol}^{-2} \text{ kg}) m_2 + (243.5 \text{ cal mol}^{-2.5} \text{ kg}^{1.5}) m_2^{1.5}$$

d) In order to obtain an expression for $\left(\dfrac{\partial \Delta H}{\partial n_1}\right)_{n_2}$, we need to express ΔH as a function of n_1. We can do that if we use the expression for the number of moles of solvent required to prepare a solution of molality m_2 containing one mole of solute. That is

$$m_2 = \frac{1 \text{ mol}}{w_2} = \frac{1 \text{ mol}}{n_1 M_1}$$

If we substitute this expression into the equation for ΔH for one mole of solute at molality m_2, we obtain

$$\Delta H = 923 \text{ cal} + (476.1 \text{ cal kg}^{0.5})(n_1 M_1)^{-0.5}$$

$$- (726.1 \text{ cal kg})(n_1 M_1)^{-1.0} + (243.5 \text{ cal kg}^{1.5})(n_1 M_1)^{-1.5}$$

and

$$L_{m1} = \left(\frac{\partial \Delta H}{\partial n_1}\right)_{n_2}$$

$$= -(0.5)(476.1 \text{ cal kg}^{0.5})M_1^{0.5} n_1^{-1.5} + (726.1 \text{ cal kg})M_1^{-1.0} n_1^{-1.0}$$

$$- (1.5)(243.5 \text{ cal kg}^{1.5})M_1^{-1.5} n_1^{-2.5}$$

If we substitute again in terms of m_2, we obtain

$$L_{m1} = -(0.5)(476.1 \text{ cal mol}^{-1.5} \text{ kg}^{0.5})M_1 m_2^{1.5} + (726.1 \text{ cal mol}^{-2.0} \text{ kg})M_1 m_2^{2.0}$$

$$- (1.5)(243.5 \text{ cal mol}^{-2.5} \text{ kg}^{1.5})M_1 m_2^{2.5}$$

$$= -(4.289 \text{ cal mol}^{-2.5} \text{ kg}^{1.5})m_2^{1.5} + (13.08 \text{ cal mol}^{-3.0} \text{ kg}^{2.0})m_2^{2.0}$$

$$- (6.582 \text{ cal mol}^{-3.5} \text{ kg}^{2.5})m_2^{2.5}$$

in good agreement with b).

16.

$$V_{m2} = \left(\frac{\partial V}{\partial n_2}\right)_{T,P,n_1} = \left(\frac{\partial V}{\partial m}\right)_{T,P,1 \text{ kg solvent}}$$

$$= 45.60 + 0.14 m$$

$$V_{m1}^{\bullet} = \frac{18.02 \text{ g mol}^{-1}}{0.988 \text{ g cm}^{-3}} = 18.24 \text{ cm}^3 \text{ mol}^{-1}$$

$$V_{m2}^{\bullet} = \frac{60.056 \text{ g mol}^{-1}}{1.335 \text{ g cm}^{-3}} = 44.49 \text{ cm}^3 \text{ mol}^{-1}$$

To compute V_{m1}, we use Equation 18-38 for the volume,

$$V_{m1} - V_{m1}^{\bullet} = \int_{V_{m2^{\bullet}}}^{V_{m2}} \frac{n_2}{n_1} dV_{m2}$$

$$= - \int_{m=0}^{m} \frac{m}{1000/18.016} 0.14 \, dm = -0.00126 \, m^2$$

$$V_{m1} = 18.24 - 0.00126 \, m^2$$

For the solution at $m = 1$, $V_{m1} = 18.24$ cm^3 mol^{-1}, $V_{m2} = 45.74$ cm^3 mol^{-1}.

a)

$$\Delta V_m = V_{m2}(m_2 = 1) - V_{m2(s)}^{\bullet}$$

$$= (45.74 - 44.99) \text{ cm}^3 \text{ mol}^{-1} = 0.75 \text{ cm}^3 \text{ mol}^{-1}$$

b) Since $V_{m1} = V_{m1}^{\bullet}$,

$$\Delta V_m = V_{m2}(m_2 = 1) - V_{m2(s)}^{\bullet}$$

$$= (45.74 - 44.99) \text{ cm}^3 \text{ mol}^{-1} = 0.75 \text{ cm}^3 \text{ mol}^{-1}$$

The calculation of V_{m1} implies that this quantity is constant within experimental error. If this result is valid, then V should be a linear function of m. Since V is given as a quadratic function of m, the data are not precise enough to detect the variation of V_{m1} with concentration.

18.   a) The plot of V^{mE} is shown in Figure 18-11.

Figure 18-11. A plot of V^{mE} as a function of X_1.

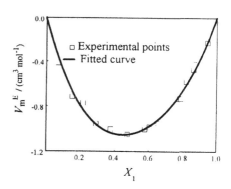

b) The best curve by nonlinear least squares is

$$V_m^E = X_1 X_2 [- 4.1902 + 0.6763(2X_1 - 1) - 1.0138(2X_1 - 1)^2]$$

The best fit curve is also shown in Figure 18-11.

20. The dilution process can be represented by the equation

$$\text{Solution}(n_1\ H_2O + n_2\ S) + n_1'\ H_2O = \text{Solution}[(n_1 + n_1'\ H_2O + n_2\}$$

where S represents the solute. The enthalpy of dilution can be expressed as

$$\Delta H_{dil} = (n_1 + n_1') H_{m1f} + n_2 H_{m2f} - n_1 H_{m1i} - n_2 H_{m2i} - n_1' H_{m1}^\bullet$$

$$= n_1 (H_{m1f} - H_{m1i}) + n_2 (H_{m2f} - H_{m2i}) + n_1' (H_{m1f} - H_{m1}^\bullet)$$

We can write for the excess enthalpy of the initial and final solutions

$$H_m^E = \Delta H_{mix} - \Delta H_{mix}^I = \Delta H_{mix}$$

$$H_{mi}^E = n_1 H_{m1i} + n_2 H_{m2i} - n_1 H_{m1}^\bullet - n_2 H_{m2}^\bullet$$

$$H_{mf}^E = (n_1 + n_1') H_{m1f} + n_2 H_{m2f} - (n_1 + n_1') H_{m1}^\bullet - n_2 H_{m2}^\bullet$$

and

$$H_{mf}^E - H_{mi}^E = n_1 (H_{m1f} - H_{m1i}) + n_1' (H_{m1f} - H_{m1}^\bullet) + n_2 (H_{m2f} - H_{m2i})$$

which is equal to ΔH_{dil}.

The problem can also be solved by constructing a thermodynamic cycle with the mixing equations for the initial and final solutions, which is equivalent to the calculation above.

22. The data given are the enthalpies of dissolving the solids in sufficient solvent that the solutes are in an infinitely dilute solution; thus, each solution process results in the same state for each solute. In order to calculate the values of H_m^E, we need to obtain the enthalpies of mixing for each solid solution. We can do that by setting up a thermodynamic cycle as follows:

$$Fe_2SiO_4(s) + solvent = Fe_2SiO_4(inf\ dil\ solution) \qquad \Delta H_1$$

$$(Mg_2SiO_4)_{X_1}(Fe_2SiO_4)_{(1-X_1)}(s) + solvent = X_1\ Mg_2SiO_4 + (1-X_1)\ Fe_2SiO_4\ (inf\ dil\ solution) \qquad \Delta H_2$$

$$Mg_2SiO_4(s) + solvent = Mg_2SiO_4(inf\ dil\ solution) \qquad \Delta H_3$$

The desired process is

$$X_1\ Mg_2SiO_4(s) + (1-X_1)\ Fe_2SiO_4(s) = (Mg_2SiO_4)_{X_1}(Fe_2SiO_4)_{(1-X_1)}(s) \qquad \Delta H_4$$

Since Equation 4 is equal to

X_1 (Equation 3) + (1 - X_1) (Equation 1) - (Equation 2)

$$\Delta H_4 = X_1\ \Delta H_3 + (1-X_1)\ \Delta H_1 - \Delta H_2$$

where H_m^E is equal to ΔH_4 for each solid. The calculated values are shown in Table 18-6.

Table 18-6.

X_1	$\Delta H_{solution}/(kJ\ mol^{-1})$	H_m^E
0.0	19.52	0
0.2	28.06	1.658
0.4	37.31	2.606
0.6	47.82	2.294
0.8	58.33	1.982
1.0	70.51	0

The Redlich-Kister equation for a regular solution is

$$H_m^E = X_1\ X_2\ [A + B\ (2\ X_1 - 1)]$$

The best nonlinear fit of the data to that equation yields $A = 10.577$ and B = 0.7083. A plot of

the experimental values of H_m^E and the best fit curve is shown in Figure 18-12.

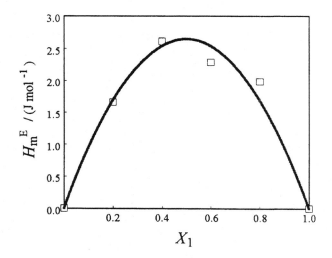

Figure 18-12. A plot of H_m^E and the best fit Redlich-Kister curve.

Chapter 19

Activity, Activity Coefficients, and Osmotic Coefficients of Strong Electrolytes

2. a)

$$K = \frac{a_{2,\text{satd}}}{a_{2,\text{pure}}} = a_{2,\text{satd}}$$

since the activity of the pure solid is equal to 1 in its standard state.

$$\therefore \quad \left(\frac{\partial \ln a_{2,\text{satd}}}{\partial T}\right)_P = \left(\frac{\partial \ln K}{\partial T}\right)_P = \frac{\Delta H_m^\circ}{RT^2}$$

$$= \frac{H_{m2,\text{satd}}^\circ - H_{m2}^\bullet}{RT^2}$$

The standard state for the solute in solution is the hypothetical one-molal standard state ; the enthalpy in that standard state is the value in the infinitely dilute solution, in which Henry's law is followed. Thus,

$$\left(\frac{\partial \ln a_{2,\text{satd}}}{\partial T}\right)_P = \frac{H_{m2}^\infty - H_{m2}^\bullet}{RT^2} = \frac{-L_{m2}^\bullet}{RT^2}$$

b) For an electrolyte that dissociates into ν particles,

$$a_2 = a_\pm^\nu = \gamma_\pm^\nu \left(m_\pm/m^\circ\right)^\nu$$

$$a_{2,\text{satd}} = \gamma_{\pm,\text{satd}}^\nu \left(m_{\pm,\text{satd}}/m^\circ\right)^\nu$$

$$\left(\frac{\partial \ln a_{2,\text{satd}}}{\partial T}\right)_P = \nu\left(\frac{\partial \ln \gamma_{\pm,\text{satd}}}{\partial T}\right)_P + \nu\left(\frac{\partial \ln m_{\pm,\text{satd}}}{\partial T}\right)_P$$

$$= -\frac{L_{m2}^\bullet}{RT^2}$$

Since,

$$m_{\pm} = [\nu_+^{\nu_+} \nu_-^{\nu_-}]^{1/\nu} \, m_2$$

$$\left(\frac{\partial \ln m_{\pm}}{\partial T}\right)_P = \left(\frac{\partial \ln m_2}{\partial T}\right)_P$$

Therefore,

$$-\frac{\overset{\bullet}{L}_{m_2}}{\nu R T^2} = \left(\frac{\partial \ln \gamma_{\pm,satd}}{\partial T}\right)_P + \left(\frac{\partial \ln m_{2,satd}}{\partial T}\right)_P$$

c) If $\ln \gamma_{\pm}$ is a function of both T and m_2, then

$$\left(\frac{\partial \ln \gamma_{\pm}}{\partial T}\right)_P = \left(\frac{\partial \ln \gamma_{\pm}}{\partial T}\right)_{P,m_2} + \left(\frac{\partial \ln \gamma_{\pm}}{\partial m_2}\right)_{P,T}\left(\frac{\partial m_2}{\partial T}\right)_P$$

For a saturated solution, we can write

$$\left(\frac{\partial \ln \gamma_{\pm,satd}}{\partial T}\right)_P = \left(\frac{\partial \ln \gamma_{\pm}}{\partial T}\right)_{P,m_2} + \left(\frac{\partial \ln \gamma_{\pm}}{\partial m_2}\right)_{P,T}\left(\frac{\partial m_{2,satd}}{\partial T}\right)_P$$

d) If we equate the expressions for $\left(\frac{\partial \gamma_{\pm,satd}}{\partial T}\right)_P$ from b) and c), we obtain

$$-\left(\frac{\partial \ln m_{2,satd}}{\partial T}\right)_P - \frac{\overset{\bullet}{L}_{m2}}{\nu R T^2} = \left(\frac{\partial \ln \gamma_{\pm}}{\partial T}\right)_{P,m_2} + \left(\frac{\partial \ln \gamma_{\pm}}{\partial m_2}\right)_{P,T}\left(\frac{\partial m_{2,satd}}{\partial T}\right)_P$$

$$\text{or} \quad -\frac{\overset{\bullet}{L}_{m2}}{\nu R T^2} = \left(\frac{\partial \ln \gamma_{\pm}}{\partial T}\right)_{P,m_2} + \left(\frac{\partial \ln \gamma_{\pm}}{\partial m_2}\right)_{P,T}\left(\frac{\partial m_{2,satd}}{\partial T}\right)_P + \left(\frac{\partial \ln m_{2,satd}}{\partial T}\right)_P$$

$$\text{But,} \quad \left(\frac{\partial \ln m_{2,satd}}{\partial T}\right)_P = \left(\frac{1}{m_{2,satd}}\right)\left(\frac{\partial m_{2,satd}}{\partial T}\right)_P$$

Thus,

$$-\frac{L_{m2}^{\bullet}}{\nu RT^2} = \left(\frac{\partial \ln \gamma_{\pm}}{\partial T}\right)_{P,m_2} + \left(\frac{\partial \ln \gamma_{\pm}}{\partial m_2}\right)_{P,T}\left(\frac{\partial m_{2,satd}}{\partial T}\right)_P + \left(\frac{1}{m_{2,satd}}\right)\left(\frac{\partial m_{2,satd}}{\partial T}\right)_P$$

$$= \left(\frac{\partial \ln \gamma_{\pm}}{\partial T}\right)_{P,m_2} + \left(\frac{\partial m_{2,satd}}{\partial T}\right)_P\left[\left(\frac{\partial \ln \gamma_{\pm}}{\partial m_2}\right)_{T,P} + \left(\frac{1}{m_{2,satd}}\right)\right]$$

e) Since the solvent does not change in the saturated solution on the addition of an infinitesimal amount of solute,

$$\Delta H_{m,soln} = H_{m2,satd} - H_{m2}^{\bullet} = L_{m2,satd} - L_{m2}^{\bullet}$$

If we substitute for $L_{m2,satd}$ from Eq. 19-75, written for a saturated solution, and for $L_{m2}^{\bullet}$ from Eq. 19-80, we obtain

$$\Delta H_{m,soln} = -\nu RT^2\left(\frac{\partial \ln \gamma_{\pm,satd}}{\partial T}\right)_P + \nu RT^2\left\{\left(\frac{\partial \ln \gamma_{\pm}}{\partial T}\right)_{P,m_2} + \left(\frac{\partial m_{2,satd}}{\partial T}\right)_P\left[\left(\frac{\partial \ln \gamma_{\pm}}{\partial m_2}\right)_{P,T} + \left(\frac{1}{m_{2,satd}}\right)\right]\right\}$$

If we now substitute for $\left(\dfrac{\partial \ln \gamma_{\pm,satd}}{\partial T}\right)_P$ from c), we obtain

$$\Delta H_{m,soln} = -\nu RT^2\left[\left(\frac{\partial \ln \gamma_{\pm}}{\partial T}\right)_{P,m_2} + \left(\frac{\partial \ln \gamma_{\pm}}{\partial m_2}\right)_{T,P}\left(\frac{\partial m_{2,satd}}{\partial T}\right)_P - \left(\frac{1}{m_{2,satd}}\right)\left(\frac{\partial m_{2,satd}}{\partial T}\right)_P - \left(\frac{\partial \ln \gamma_{\pm}}{\partial T}\right)_{P,m_2} - \left(\frac{\partial \ln \gamma_{\pm}}{\partial m_2}\right)_{T,P}\left(\frac{\partial m_{2,satd}}{\partial T}\right)_P\right]$$

$$= \frac{\nu RT^2}{m_{2,satd}}\left(\frac{\partial m_{2,satd}}{\partial T}\right)_P$$

f) If the solute is a nonelectrolyte, $\nu = 1$, and

$$\Delta H_{m,soln} = \frac{RT^2}{m_{2,satd}}\left(\frac{\partial m_{2,satd}}{\partial T}\right)_P$$

4. When the barometric temperature is 23.8°C and the barometer reading is 751.0 mm Hg on a brass scale, the brass scale correction for the thermal expansion of Hg from 0°C (CRC Handbook) is -2.9 mm Hg. Therefore, the atmospheric pressure is 751.0 mm Hg - 2.9 mm Hg, or 748.1 mm Hg.

The pressure of hydrogen that is bubbled through a column of aqueous solution 0.68 cm high is higher than the atmospheric pressure by the hydrostatic pressure of that column.

$$\Delta P = h\rho g = (0.68 \times 10^{-2}\ m)(0.997 \times 10^{3}\ kg\ m^{-3})(9.8\ m\ s^{-2})$$

$$= 66.4\ Pa$$

$$P_{atm} = (748.1\ mm\ Hg)[133.32\ Pa\ (mm\ Hg)^{-1}]$$

$$= 99740\ Pa$$

$$P_{H_2} = (99740 + 66.4)\ Pa = 99806\ Pa = 0.9981\ Bar$$

For the cell:

$$H_2(g),\ HCl(m_2 = 0.002951\ mol\ kg^{-1}),\ AgCl(s),\ Ag(s)$$

the cell reaction is

$$\tfrac{1}{2}\ H_2(g,\ 0.9981\ Bar) + AgCl(s) = H^+(m_+ = 0.002951) + Cl^-(m_- = 0.002951) + Ag(s)$$

and the Nernst equation is

$$\mathscr{E} = \mathscr{E}^\circ - \frac{RT}{\mathscr{F}}\ \ln\ \frac{a_H \cdot a_{Cl^-}}{(p_{H_2}/p^\circ)^{0.5}}$$

$$= \mathscr{E}^\circ - \frac{RT}{\mathscr{F}}\ \ln\ \frac{\left[(m_H \cdot m_{Cl^-})/(m^\circ)^2 \right]\gamma_\pm^2}{(p_{H_2}/p^\circ)^{0.5}}$$

The accepted value for $\mathscr{E}^\circ$ is 0.22228 V; we can now solve for $\gamma_\pm$.

$$\ln\ \gamma_\pm^2 = \mathscr{E}^\circ - \mathscr{E} - \frac{RT}{\mathscr{F}}\ \ln\ \frac{(m_H \cdot m_{Cl^-})/(m^\circ)^2}{(p_{H_2}/p^\circ)^{0.5}}$$

$$= 0.22228 \text{ V} - 0.52393 \text{ V}$$

$$- \frac{(8.3145 \text{ J mol}^{-1} \text{ K}^{-1})(298.15 \text{ K})}{96485 \text{ C mol}^{-1}} \ln \frac{\left(\dfrac{0.002951 \text{ mol kg}^{-1}}{1.000 \text{ mol kg}^{-1}}\right)^2}{\left(\dfrac{0.9981 \text{ Bar}}{1.0000 \text{ Bar}}\right)^{0.5}}$$

$$= -2.32 \times 10^{-3}$$

$$\gamma_\pm^2 = 0.99768 \qquad \gamma_\pm = 0.99884$$

With the value of $\gamma_\pm$, we can calculate the potential at a pressure of hydrogen of 1.01325 Bar.

$$\mathscr{E} = \mathscr{E}^\circ - \frac{RT}{\mathscr{F}} \ln \frac{\left[(m_H \cdot m_{Cl^-})/(m^\circ)^2\right]\gamma_\pm^2}{\left(p_{H_2}/p^\circ\right)^{0.5}}$$

$$= 0.22228 \text{ V} - \frac{(8.3145 \text{ J mol}^{-1} \text{ K}^{-1})(298.15 \text{ K})}{96485 \text{ C mol}^{-1}} \ln \frac{(0.002951)^2(0.99768)}{(1.01325)^{0.5}}$$

$$= 0.52186 \text{V}$$

6.   a) In Figure 19-1, we have plotted the solubility of silver chloride against the concentration of

supporting electrolyte for KNO_3 and $NaNo_3$, and against three times the total concentration for

$Ba(NO_3)$ in the lower set of points, . In the upper set of points, we have plotted the solubility of

AgCl against the total concentration of $Ba(NO_3)_2$. It can be seen that all the points in the lower

set fall along the same curve, whereas the solubility plotted against the total concentration of

$Ba(NO_3)_2$ fall on a different curve. In the solutions other than $Ba(NO_3)_2$, the ionic strength is

identical to the total concentration of electrolyte. For $Ba(NO_3)_2$, the ionic strength is three times

the total concentration.

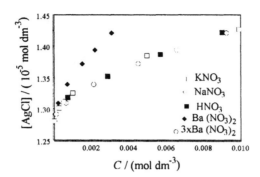

Figure 19-1. The solubility of AgCl as a function of the concentration of supporting electrolyte.

b) These results are in accordance with the principle that the properties of electrolytes depend on the ionic strength and not the total concentration of electrolyte.

c)d) The concentration of AgCl is 1.385×10^{-5} mol L^{-1} in a 0.004972 M solution of KNO$_3$. Therefore,

$$I = 0.5 \ \Sigma \ c_1 z_i^2$$

$$= 0.5\,[(0.004972)(1)^2 \ + \ (0.004972)(1)^2 \ + \ (1.38 \times 10^{-5})(1)^2 \ + \ (1.38 \times 10^{-5})(1)^2]$$

$$= 0.004986$$

The concentration of AgCl is 1.4212×10^{-5} mol L^{-1} in a 0.003083 M solution of Ba(NO$_3$)$_2$. Therefore,

$$I = 0.5\,[(0.003083)(2)^2 \ + \ (2)(0.003083)(1)^2 \ + \ (2)(1.421 \times 10^{-5})(1)^2]$$

$$= 0.009263$$

The quantity - log [AgCl] is plotted against $I^{0.5}$ in Figure 19-2 for all the data, indicating again that the principle of ionic strength is followed. The solid line is fitted to the data in Ba(NO$_3$)$_2$ solution. The dashed line is the Debye-Hückel limiting law. The slope of the fitted line is -0.4913; the slope of the Debye-Hückel line is 0.5091.

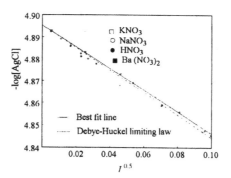

Figure 19-2. A plot of the negative logarithm of the solubility of AgCl as a function of $I^{0.5}$.

e) The activity of AgCl in the absence of supporting electrolyte is obtained from the intercept of the fitted line, 4.8953.

$$- \log \{[AgCl]/c^\circ\} = -4.8953$$

$$[AgCl]/c^\circ = 1.273 \times 10^{-5}$$

$$a_{AgCl} = \{[AgCL]/c^\circ\}^2 = 1.62 \times 10^{-10}$$

The activity of AgCl is the same at all ionic strengths, since the solution is in equilibrium with pure solid at all ionic strengths. The concentrations and activity coefficients vary with ionic strength.

f) The solubility product is equal to the activity of AgCl, 1.620×10^{-10}.

8.   a) The cell reaction is

$$H^+ (m = 0.001) + Cl^- (m = .001) = H^+ (m = 0.0001) + Cl^- (m = .0001)$$

$$\mathscr{E} = -\frac{RT}{\mathscr{F}} \ln \frac{a_{H^+}(m = 0.0001)a_{Cl^-}(m = 0.0001)}{a_{H^+}(m = 0.001)a_{Cl^-}(m = 0.001)}$$

$$= -\frac{RT}{\mathscr{F}} \ln \frac{(0.0001)^2 \gamma_{\pm(0.0001)}^2}{(0.001)^2 \gamma_{\pm(0.001)}^2}$$

In 0.001 molal HCl,

$$\log \gamma_\pm = -0.5091\,m^{0.5} = -0.5091(0.001)^{0.5} = -0.016099$$

$$\gamma_\pm = 0.9636$$

In 0.0001 molal HCl,

$$\log \gamma_\pm = -0.005091$$

$$\gamma_\pm = 0.9883$$

$$\mathscr{E} = -\frac{(8.3145 \text{ J mol}^{-1}\text{ K}^{-1})(298.15 \text{ K})}{96485 \text{ C mol}^{-1}} \ln \frac{(0.0001)^2(0.9883)^2}{(0.001)^2(0.9636)^2}$$

$$= 0.1170\text{V}$$

A positive potential is consistent with the spontaneous transfer of solute from a more concentrated

solution to a more dilute solution.

b) The cell reaction is

$$Mg^{++}\ (m = 0.0001) + SO_4^=\ (m = 0.0001) = Mg^{++}\ (m = 0.001) + SO_4^=\ (m = 0.001)$$

$$\mathscr{E} = -\frac{RT}{\mathscr{F}} \ln \frac{(0.001)^2 \gamma_{\pm(0.001)}^2}{(0.0001)^2 \gamma_{\pm(0.0001)}^2}$$

In 0.001 molal $MgSO_4$,

$$\log \gamma_\pm = -0.5091\,(2)^2[4\,(0.001)]^{0.5}$$

$$= -0.2188$$

$$\gamma_{\pm} = 0.7434$$

In 0.0001 molal $MgSO_4$,

$$\log\gamma_{\pm} = -0.04073$$

$$\gamma_{\pm} = 0.9105$$

$$\mathscr{E} = -\frac{(8.3145 \text{ J mol}^{-1} \text{ K}^{-1})(298.15 \text{ K})}{2(96485 \text{ C mol}^{-1})} \ln \frac{(0.001)^2(0.734)^2}{(0.0001)^2(0.9105)^2}$$

$$= -0.05395 \text{ V}$$

The negative potential is consistent with the nonspontaneous transfer of solute from a more dilute solution to a more concentrated solution. The difference in magnitude between the results in a) and b) reflects the difference in activity coefficients in $MgSO_4$ and HCl solutions of the same molality, as well as the factor of two for the number of electrons transferred in b).

10.   Table 19-1 shows Keefer's data and calculated values of I, $I^{0.5}$, the product $\left[m_{Cu^{++}}\left(m_{IO_3^-}\right)^2\right]$, and the logarithm of the product.

a) Figure 19-3 shows the logarithm of the solubility product (column 6) plotted against $I^{0.5}$. The curve is fitted by nonlinear fitting to the function

$$\log \left(m_{Cu^{++}}\right)\left(m_{IO_3^-}\right)^2 = -7.1376 + \frac{3.077 I^{0.5}}{1 + 1.13 I^{0.5}}$$

The extrapolated value is -7.1376, so that K_{SP} is 7.29×10^{-8}.

b) From the fitted equation, $Ba_i = 1.13$ and from Table 19-4, $B = 3.285\times10^{-9}$, so that $a_i = 2.91\times10^{-9}$ m.

Table 19-1. Calculations for the solubility of $Cu(IO_3)_2$.

$m_{KCl}/(\text{mol kg}^{-1})$	$m_{Cu(IO_3)_2}/(\text{mol kg}^{-1})$	I	$I^{0.5}$	$m_{Cu^{++}}\, m_{IO_3^-}^2 \times 10^4$	$\log m_{Cu^{++}}\, m_{IO_3^-}^2$
0.00000	0.003245	0.00974	0.09867	0.001367	-6.864
0.00501	0.003398	0.01520	0.12330	0.001569	-6.804
0.01002	0.003517	0.02057	0.14343	0.001740	-6.759
0.02005	0.003730	0.03124	0.17675	0.002076	-6.683
0.03511	0.003975	0.04704	0.21688	0.002512	-6.600
0.05017	0.004166	0.06267	0.25034	0.002892	-6.539
0.07529	0.004453	0.08865	0.29774	0.003532	-6.452
0.10050	0.004694	0.11458	0.33850	0.004137	-6.383

Figure 19-3. A plot of the logarithm of the solubility product against $I^{0.5}$.

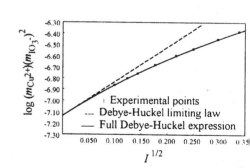

c)

$$\log K = \log \left[a_{Cu^{++}} a_{IO_3^-}^2 \right]$$

$$= \log \left[(m_{Cu^{++}}/m^\circ)(\gamma_{Cu^{++}})(m_{IO_3^-}/m^\circ)^2 (\gamma_{IO_3^-})^2 \right]$$

$$= \left[(m_{Cu^{++}}/m^\circ)(m_{IO_3^-}/m^\circ)^2 (\gamma_\pm)^3 \right]$$

$$\log \left[(m_{Cu^{++}}/m^\circ)(m_{IO_3^-}/m^\circ)^2 \right] = \log K - 3 \log \gamma_\pm$$

$$= \log K - 3 \frac{-A(1)(2)I^{0.5}}{1 + Ba_i I^{0.5}}.$$

Thus, $\log K$ is the constant in the fitted equation.

d) See graph in a)

Chapter 20

Changes in Gibbs Function for Processes Involving Solutions

2. $\Delta_f G_{m,298K}^\circ = -394.359$ kJ mol^{-1} for $CO_2(g)$.

For the reaction

$$CO_2(g) = CO_2(aq)$$

$$\Delta G_m^\circ = \mu^\circ[CO_2 aq)] - \mu^\circ[CO_2(g)]$$

$$= \mu^\circ[CO_2(g)] + RT \ln (k_2''/p_{gas}^\circ) - \mu^\circ[CO_2(g)]$$

$$= RT \ln (k_2''/f^\circ)$$

The value of k_2'' for $CO_2(g)$ dissolved in water at 298 K is 29.5 atm. Thus,

$$\Delta G_m^\circ = (8.3145 \text{ J mol}^{-1} \text{ K}^{-1})(298.15 \text{ K}) \ln (29.5 \text{ atm})/(0.986 \text{ atm})$$

$$= 8.424 \text{ kJ mol}^{-1}$$

$$CO_2(aq) + H_2O(\ell) = HCO_3^-(aq) + H^+(aq)$$

$$K_1 = 3.5 \times 10^{-7}$$

$$\Delta G_{m,298K}^\circ = -(8.3145 \text{ J mol}^{-1} \text{ K}^{-1})(298.15 \text{ K}) \ln (3.5 \times 10^{-7})$$

$$= 36.85 \text{ kJ mol}^{-1}$$

$$HCO_3^-(aq) + H_2O(\ell) = CO_3^=(aq) + H^+(aq)$$

$$K_2 = 3.7 \times 10^{-11}$$

$$\Delta G_{m,298K}^\circ = -(8.3145 \text{ J mol}^{-1} \text{ K}^{-1})(298.15) \ln (3.7 \times 10^{-11})$$

$$= 59.55 \text{ kJ mol}^{-1}$$

$$= \Delta G_{m,298K}{}^{\circ}\{[CO_3{}^{=}(aq) + [H^+(aq)] - [HCO_3{}^-(aq)]\}$$

Thus,

$$\Delta_f G_{m,298.15K}{}^{\circ}[CO_3{}^{=}(aq)] = 59.55 \text{ kJ mol}^{-1} - \Delta_f G_m{}^{\circ}[H^+(aq)] + \Delta_f G_m{}^{\circ}[HCO_3{}^-(aq)]$$

$$= 59.55 \text{ kJ mol}^{-1} - 0 + \Delta_f G_m{}^{\circ}[HCO_3{}^-(aq)]$$

From K_1,

$$36.85 \text{ kJ mol}^{-1} = \Delta_f G_{m,298.15K}{}^{\circ}\{[HCO_3{}^-(aq)] + [H^+(aq)] - [Co_2(aq)] - [H_2O(\ell)]\}$$

$$\Delta_f G_{m,298.15K}{}^{\circ}[HCO_3{}^-(aq)] = 36.85 \text{ kJ mol}^{-1} + \Delta_f G_{m,298.15K}{}^{\circ}\{[Co_2(aq)] + [H_2O(\ell)] - [H^+(aq)]\}$$

$$= (36.85 - 394.359 + 8.424 - 237.129) \text{ kJ mol}^{-1}$$

$$= -586.21 \text{ kJ mol}^{-1} \qquad \text{and,}$$

$$\Delta_f G_{m,298.15K}{}^{\circ}[CO_3{}^{=}(aq)] = (59.55 - 586.21) \text{ kJ mol}^{-1} = -526.66 \text{ kJ mol}^{-1}$$

4.   a) α-D-glucose(s) = β-D-glucose(s)   (*i*)

Since we have no data that bear directly on this equilibrium, we need to find a series of reactions that sum to Reaction (*i*).

Since α-D-glucose(s) and β-D-glucose(s) follow Henry's law (activity is proportional to concentration), for both species,

$$\mu(\text{solute}) = \mu^{\circ}(\text{solute}) + RT \ln\frac{c_2}{c^{\circ}}$$

and, for the saturated solutions,

$$\mu(\text{solid}) = \mu^{\circ}(\text{solid}) = \mu^{\circ}(\text{solute}) + RT \ln \left(\frac{c_2}{c^{\circ}}\right)_{satd}$$

Therefore, for the Reaction (*ii*)

$$\alpha\text{-D-glucose(s)} = \alpha\text{-D-glucose(solute)} \qquad (ii)$$

$$\Delta G_m^{\circ}(ii) = \mu^{\circ}(\text{solute}) - \mu^{\circ}(\text{solid})$$

$$= - RT \ln \left(\frac{c_2}{c^{\circ}} \right)_{\text{satd}}$$

$$= - (8.3145 \text{ J mol}^{-1} \text{ K}^{-1})(293.15\text{K}) \ln \frac{(20\text{g L}^{-1})/(180.15 \text{ g mol}^{-1})}{1 \text{ mol L}^{-1}}$$

$$= 5.36 \text{ kJ mol}^{-1}$$

If neither solute has any influence on the chemical potential of the other, the solubilities are independent of the presence of the other solute. Then, the additional 25 g L^{-1} when β-D-glucose is formed in a solution of α-D-glucose must be β-D-glucose. Therefore, the equilibrium constant for Reaction (*iii*)

$$\alpha\text{-D-glucose(solute)} = \beta\text{-D-glucose(solute)} \qquad (iii)$$

is

$$K = \frac{(c/c^{\circ})_{\beta}}{(c/c^{\circ})_{\alpha}} = \frac{(25 \text{ g L}^{-1})/(180.15 \text{ g mol}^{-1})}{(20 \text{ g L}^{-1})/(180.15 \text{ g mol}^{-1})}$$

$$= 1.25$$

$$\Delta G_m^{\circ}(iii) = - RT \ln K$$

$$= - (8.3145 \text{ J mol}^{-1} \text{ K}^{-1})(293.15\text{K}) \ln (1.25)$$

$$= - 0.54 \text{ kJ mol}^{-1}$$

For reaction (*iv*)

$$\beta\text{-D-glucose(solid)} = \beta\text{-D-glucose(solute)} \qquad (iv)$$

as for Reaction (*ii*)

$$\Delta G_m^\circ (iv) = \mu^\circ (solute) - \mu^\circ (solid)$$

$$= - RT \ln \left(\frac{c_2}{c^\circ} \right)_{satd}$$

$$= - (8.3145 \text{ J mol}^{-1} \text{ K}^{-1})(293.15\text{K}) \ln \frac{(49 \text{ g L}^{-1})/(190.15 \text{ g mol}^{-1})}{1 \text{ mol L}^{-1}}$$

$$= 3.17 \text{ kJ mol}^{-1}$$

Reaction (*i*) is equal to Reaction (*ii*) + Reaction (*iii*) - Reaction (*iv*), so that

$$\Delta G_m^\circ (i) = (5.36 - 0.54 - 3.17) \text{ kJ mol}^{-1} = 1.65 \text{ kJ mol}^{-1}$$

b) For Reaction (*iii*)

$$K = 1.25 \qquad \text{and}$$

$$C_2(\alpha\text{-D-glucose}) = \frac{C_2(\beta\text{-D-glucose})}{1.25}$$

$$= \frac{(49 \text{ g L}^{-1})/(180.15 \text{ g mol}^{-1})}{1.25}$$

$$= 0.217 \text{ mol L}^{-1}$$

$$T_\beta = 49 \text{ g L}^{-1} + (0.217 \text{ mol L}^{-1})(180.15 :g \text{ mol}^{-1})$$

$$= (49 + 39) \text{ g L}^{-1}$$

The concentration of a saturated solution of β-D-glucose is 49 g L^{-1}. The additional 39 g L^{-1} of α-D-glucose is greater than its solubility. Thus, this system is unstable thermodynamically. Solid α-D-glucose would precipitate out of solution until its equilibrium solubility is reached. Then Reaction (*iii*) would no longer be in equilibrium, and more α-D-glucose would be formed. The process would continue until the system consists only of solid α-D-glucose and the solution of α-D-glucose and β-D-glucose.

6.

$$\Delta G_m^{\circ} = - RT \ln K$$

An error of a factor of two in K would result in an error of $\ln 2 = 0.69$ in $\ln K$ and an error of $(8.3145 \text{ J mol}^{-1} \text{ K}^{-1})(298.15 \text{ K})(0.69) = 1.7 \text{ kJ mol}^{-1}$ in ΔG_m°.

8.   Since A' is 1000 J mol^{-1} higher in G than A'', the former will have a more negative or less positive free energy of solution than the latter. Thus A' is more soluble than A''. In order to calculate the ratio of solubilities, we consider three processes:

$$A'(s) = A''(s) \qquad (1)$$

$$\Delta G_m^{\circ}(1) = - 1000 \text{ J mol}^{-1} = \mu^{\circ}[A''(s)] - \mu^{\circ}[A'(s)]$$
$$A'(s) = A(\text{solute, satd}) \qquad (2)$$

At equilibrium $\mu^{\circ}[A'(s)] = \mu^{\circ}[A, \text{solute}] + RT \ln \dfrac{C'[A, \text{solute}]}{C^{\circ}}$

$$A''(s) = A(\text{solute, satd}) \qquad (3)$$

At equilibrium $\mu^{\circ}[A''(s)] = \mu^{\circ}[A, \text{solute}] + RT \ln \dfrac{C''(A, \text{solute})}{C^{\circ}}$

If we subtract Equation 2 from Equation 3, we obtain

$$\mu^{\circ}[A''(s)] - \mu^{\circ}[A'(s)] = RT \ln \dfrac{C''}{C'} = \Delta G_m^{\circ}(1)$$

Therefore,

$$\ln \dfrac{C''(A'', \text{solute,satd})}{C'(A', \text{solute,satd})} = \dfrac{\Delta G_m^{\circ}(1)}{RT}$$

$$= - \dfrac{1000 \text{ J mol}^{-1}}{(8.3145 \text{ J mol}^{-1} \text{ K}^{-1})(298.15 \text{ K})}$$

$$= - 0.04034$$

$$\dfrac{C''}{C'} = 0.668$$

10.   a)

$$\left(\frac{\partial \ln \gamma_\pm}{\partial T}\right)_{P,m_2} = -\frac{L_{m2}}{\nu RT^2}$$

$$= -\frac{188 \text{ J mol}^{-1}}{(2)(8.3145 \text{ J mol}^{-1} \text{ K}^{-1})T^2} = -\frac{11.13 \text{ K}}{T^2}$$

$$\ln \frac{\gamma_\pm(273 \text{ K})}{\gamma_\pm(298 \text{ K})} = 11.13 \text{ K}\left[\frac{1}{273} - \frac{1}{298}\right] = 3.42 \times 10^{-3}$$

$$[\gamma_\pm(273 \text{ K})]/[\gamma_\pm(298 \text{ K})] = 1.003$$

$$\gamma_\pm(273 \text{ K}) = (1.003)(0.89) = 0.893$$

b)

$$\log \gamma_\pm = -(0.4904)(1)(1)(0.01)^{0.5} = -4.904 \times 10^{-2}$$

$$\gamma_\pm = 0.892$$

This is good agreement.

12.   The cell reaction is

$$H^+(0.01) + Cl^-(0.01) = H^+[0.001, KNO_3(0.009)] + Cl^-[0.001, KNO_3(0.009)]$$

$$\mathscr{E} = -\frac{RT}{\mathscr{F}} \ln \frac{m_\pm^2 \gamma_\pm^2 [KNO_3(0.009)]}{m_\pm^2 \gamma_\pm^2 (0.01)}$$

But $\gamma_\pm$ is the same in both solutions because they both have the same ionic strength. Thus,

$$\mathscr{E} = -\frac{RT}{n\mathscr{F}} \ln \frac{(0.001)^2}{(0.01)^2}$$

$$= -\frac{2(8.3145 \text{ J mol}^{-1} \text{ K}^{-1})(298.15 \text{ K})}{96485 \text{ C mol}^{-1}} \ln (0.1)$$

$$= 0.1183 \text{ V}$$

14.   The cell reaction is

$$Tl^+(satd) + Cl^-(satd) = Tl^+(satd) + Cl^-(satd)$$

Both electrolyte solutions are in equilibrium with solid TlCl, so that the net reaction is zero, and

the chemical potential of TlCl is the same in both solutions. Thus $\Delta G_m = 0$ for the cell pair, and

$\mathcal{E} = 0$.

16.   In both solutions, the controlling equilibrium is

$$HOAc = H^+ + OAc^- \qquad and$$

$$K_a = \frac{c_{H^+} c_{OAc^-} \gamma_\pm^2}{a_{HOAc}}$$

Since the ionic strength is the same in both solutions, $\gamma_\pm$ is the same in both solutions. Then,

$$\frac{a_{HOAc}(1)}{a_{HOAc}(2)} = \frac{c_{H^+}(1) c_{OAc^-}(1)}{c_{H^+}(2) c_{OAc^-}(2)}$$

In solution 1,

$$c_{OAc^-} = 0.0004$$

From the value of the ionic strength,

$$0.0100 = c_{H^+} + c_{OAc^-} + c_{Cl^-}$$

$$c_{H^+} = 0.0100 - c_{Cl^-} - 0.0004 = 0.0096 - c_{Cl^-}$$

From the electrical neutrality of the solution,

$$c_{H^+} = c_{Cl^-} + c_{OAc^-} = 0.0004 + c_{Cl^-}$$

If we solve the two equations for two unknowns, we find

$$c_{H^+} = 0.0050$$

In solution 2, by the same procedure,

$$c_{H^+} = 0.0050$$

Therefore,

$$\frac{a_{HOAc}(1)}{a_{HOAc}(2)} = \frac{(0.0050)(0.0004)}{(0.0050)(0.0001)} = 4$$

18.   a) The solubility in 0.0005 molar KNO_3 is the same as in a $MgSO_4$ solution of the same ionic

strength. Thus,

$$0.0005 = 0.5[c_{Mg^{++}}(4) + c_{SO_4^=}(4)] = 4\, c_{MgSO_4}$$

$$c_{MgSO_4} = 0.000125$$

b) The equilibrium for the solubility reaction, the solubility product constant, is

$$K = c_+ c_- \gamma_{\pm}^2$$

Since the solubility is 3.21×10^{-5} M in a 0.0005 M KNO_3 solution,

$$I = 0.0005 + 9(3.251 \times 10^{-5})^2 = 0.000793$$

$$\log \gamma_{\pm} = -(0.5091)(3)(3)(0.000793)^{0.5} = -0.1290$$

$$\gamma_{\pm} = 0.7430$$

$$K = (3.251 \times 10^{-5})^2 (0.7430)^2 = 5.835 \times 10^{-10}$$

We will need to use an iterative procedure (which can be programmed for a computer) to obtain

$\gamma_{\pm}$ since we do not know the ionic strength until we know the solubility. Let us assume $\gamma_{\pm}$ is equal

to 1. Then

$$S_1 = K^{0.5} = 2.416 \times 10^{-5}$$

$$I = 9(2.416 \times 10^{-5}) = 2.174 \times 10^{-4}$$

$$\log \gamma_{\pm} = -(0.5091)(3)(3)(2.174 \times 10^{-4})^{0.5} = -0.06756$$

$$\gamma_{\pm} = 0.8559$$

$$S_2 = [K/(\gamma_\pm)^2]^{0.5} = [(5.835 \times 10^{-10})/(0.8559)^2]^{0.5} = 2.822 \times 10^{-5}$$

$$I = 2.540 \times 10^{-4}$$

$$\log \gamma_\pm = -0.07302$$

$$\gamma_\pm = 0.8452$$

$$S_3 = 2.859 \times 10^{-5}$$

$$I = 2.573 \times 10^{-4}$$

$$\log \gamma_\pm = -0.07349$$

$$\gamma_\pm = 0.8443$$

$$S_4 = 2.861 \times 10^{-5}$$

This result is now constant within experimental error.

c) If we neglect the contribution of the complex salt to the ionic strength,

$$I = (4)(0.0025) = 0.0100$$

$$\log \gamma_\pm = -(0.5091)(3)(3)(0.0100)^{0.5} = -0.4582$$

$$\gamma_\pm = 0.3482$$

$$S = [K/(\gamma_\pm)^2]^{0.5} = 6.937 \times 10^{-5}$$

The contribution of the complex salt to the ionic strength is

$$9(6.937 \times 10^{-5}) = 6.224 \times 10^{-4}, \text{ and}$$

$$I = 0.01062$$

$$\log \gamma_\pm = -0.4722$$

$$\gamma_\pm = 0.3371$$

$$S = 7.157 \times 10^{-5}$$

This error is not negligible. Another iteration yields

$$I = 0.1064$$

$$\log \gamma_{\pm} = -0.4726$$

$$\gamma_{\pm} = 0.3368$$

$$S = 7.172 \times 10^{-5}$$

which is reasonable agreement.

d) Let S = molarity of complex salt in saturated solution

C = molarity of NaCL

$$S + C = 0.0049$$

$$S = [K/(\gamma_{\pm})^2]$$

An exact expression for $\log \gamma_{\pm}$ is

$$\log \gamma_{\pm} = -(0.5091)(3)(3)(C + 9S)^{0.5}$$

but we do not know either C or S. We will therefore use an iterative procedure, starting with an

assumed value of $I = 0.0049$. Then

$$\log \gamma_{\pm} = -(0.5091)(3)(3)(0.0049)^{0.5} = -0.3207$$

$$\gamma_{\pm} = 0.4778$$

$$S = 5.056 \times 10^{-5}$$

$$C = 0,0049 - 5.056 \times 10^{-5} = 0.004849$$

$$I = C + 9S = 0.005304$$

$$\log \gamma_{\pm} = -(0.5091)(3)(3)(0.005304)^{0.5} = -0.3337$$

$$\gamma_{\pm} = 0.4638$$

$$S = 5.208 \times 10{-5}$$

$$C = 0.0049 - 5.208 \times 10{-5} = 0.004848$$

$$I = 0.005317$$

$$\log \gamma_\pm = -0.3341$$

$$\gamma_\pm = 0.4634$$

$$S = 5.213 \times 10^{-5}, \text{ which is close enough.}$$

$$C = 0.0049 - 5.213 \times 10^{-5} = 0.004848$$

20. $$S = 0.00087$$

$$I = 4(0.00087) = 0.003480$$

$$\log \gamma_\pm = -(0.5091)(2)(2)(0.003480)^{0.5} = -0.1201$$

$$\gamma_\pm = 0.7583$$

$$K = S^2 \gamma_\pm^2 = 4.35 \times 10^{-7}$$

$$\Delta G_m^\circ = -RT \ln K$$

$$= -(8.3145 \text{ J mol}^{-1} \text{ K}^{-1})(298.15 \text{ K}) \ln (4.35 \times 10^{-7})$$

$$= 36.3 \text{ kJ mol}^{-1}$$

Chapter 21

Systems Subject to a Gravitational Field

2.   At constant temperature,

$$dG_{m2} = \left(\frac{\partial G_{m2}}{\partial \ln m}\right)_{P,x} d \ln m + \left(\frac{\partial G_{m2}}{\partial P}\right)_{x,\ln m} dP + \left(\frac{\partial G_{m2}}{\partial x}\right)_{P,\ln m} dx$$

$$= RT\, d\,(\ln m) + V_{m2} dP - M_2 g dx = 0$$

From Equation 21-11,

$$dP = -\rho g dx$$

Therefore, we can write

$$RT\, d\,(\ln m) = M_2 g dx - M_2 v_2 \rho g dx = M_2 g(1 - v_2 \rho) dx$$

$$RT\int_{m_{surface}}^{m_d} d \ln m = M_2 g(1 - v_2 \rho)\int_0^d dx$$

$$RT \ln \frac{m_d}{m_{surface}} = M_2 g(1 - v_2 \rho) d$$

$$\ln \frac{m_d}{m_{surface}} = \frac{M_2 g}{RT}(1 - v_2 \rho) d$$

$$= \frac{(2.016\times10^{-3}\text{ kg mol}^{-1})(9.8\text{ m s}^{-2})[1 - (13.0\times10^{-3}\text{ m}^3\text{ kg}^{-1})(0.998\times10^3\text{ kg m}^{-3}]}{(8.3145\text{ J mol}^{-1}\text{ K}^{-1})(298.15\text{K})}$$

$$= -(9.5\times10^{-5}\text{ m}^{-1})d$$

4.   If the solute is monodisperse, Equation 21-34 says that $\ln c'$ should be a linear function of x^2.

Table 21-1, taken from a spreadsheet, shows the calculated values.

Table 21-1.

$x/(10^{-2}$ m)	$c''/(g\ dm^{-3})$	$x^2/(10^{-4}m^2)$	$\ln\ [c''/(g\ dm^{-3})]$
4.90	1.30	24.0	0.262
4.95	1.46	24.5	0.378
5.00	1.64	25.0	0.495
5.05	1.84	25.5	0.610
5.10	2.06	26.0	0.723
5.15	2.31	26.5	0.837

The data are plotted in Figure 21-1, together with a line fitted by the method of least squares.

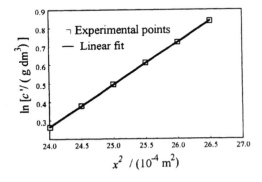

Figure 21-1. Data from Table 21-1, together with the best fit line.

The equation of the best-fit line is

$$\ln [c/(\text{g dm}^{-3})] = -5.270 + (2305 \text{ m}^{-2}) x^2$$

According to Equation 21-34, the slope should be given by

$$\text{slope} = \frac{\omega^2}{2RT} M_2 (1 - v_2 \rho) \quad \text{and}$$

$$M_2 = \frac{2RT(\text{slope})}{\omega^2 (1 - v_2 \rho)}$$

$$= \frac{(2)(8.3145 \text{ J mol}^{-1} \text{ K}^{-1})(293.15 \text{ K})(2305 \text{ m}^{-2})}{[(2)(3.1416)(182.8 \text{ s}^{-1})]^2 [1 - (0.7514 \text{ dm}^3 \text{ kg}^{-1})(1.034 \text{ kg dm}^{-3})]}$$

$$= 38.1 \text{ kg mol}^{-1}$$

Chapter 22

Estimation of Thermodynamic Quantities

2. a) By the method of Watson, Anderson, Beyer, and Yoneda, for 1,2-dibromoethane(g)

$$S_m^{\circ}$$

Base group, methane	186.19 J mol^{-1} K^{-1}
Primary CH$_3$	43.30 J mol^{-1} K^{-1}

$\rightarrow$CH$_3$-CH$_3$

To provide two CH$_3$ groups for substitution by Br

Two primary	2×43.30 J mol^{-1} K^{-1}
Substitution of 2 Br atoms for 2 CH$_3$ groups	2×13.10 J mol^{-1} K^{-1}
Two type two	4× - 5.23 J mol^{-1} K^{-1}
SUM	321.37 J mol^{-1} K^{-1}

The NBS tables give $S_m^{\circ} = 333.1$ J mol^{-1} K^{-1}. The agreement is only fair.

b) Vapor pressure data from Boublík *et al* does not extend below 52°C, and the NIST WebBook used the Antoine equation, so we used an empirical equation from C. L. Yaws, Handbook of Vapor Pressure, Gulf Publishing Company, Houston, 1994, p. 343,

$$\ln (p/Pa) = 43.7786 - \frac{5.5876 \times 10^3}{T} - 3.0891 \ln T$$

$$- 1.3834 \times 10^{-9} T + 8.2665 \times 10^{-7} T^2$$

as translated to our choice of units. We used this equation to calculate $\ln p$ as a function of T over the range from 283.15 K tp 328.15 K. We obtained a value of $\ln p$ equal to 7.5107 at 298.15 K, or $p = 1828$ Pa. Then

$$\Delta S_m = \int_{0.1 \ MPa}^{1827 \ Pa} \left(\frac{\partial S_m}{\partial P} \right)_T dP = \int_{0.1 \ MPa}^{1827 \ Pa} - \left(\frac{\partial V_m}{\partial T} \right)_P dP$$

$$= \int_{0.1 \ MPa}^{1827 \ Pa} - \frac{R}{P} dP = - R \ln \left(\frac{1828 \ Pa}{0.1 \ MPa} \right)$$

$$= 33.27 \ \text{J mol}^{-1} \ \text{K}^{-1}$$

c) We can use the data from the same empirical equation to obtain an equation for $d \ln p/ dt$, and and calculate $(d \ln p/dT)$ from that equation. The equation is

$$\frac{d \ln p}{dT} = \frac{5.587 \times 10^3}{T^2} - \frac{3.0891}{T} - 1.3834 \times 10^{-9} + 16.5330 \, T$$

The value obtained at 298.15 K is 0.052989 K^{-1}. The value of ΔH_m was calculated as $RT^2(d \ln p/dT)$. The value at 298.15 K is 39.16 kJ mol^{-1}.

d) From c)

$$\Delta S_m(\text{vap}) = \frac{39.16 \ \text{kJ mol}^{-1}}{298.15 \ \text{K}} = 131.3 \ \text{J mol}^{-1} \ \text{K}^{-1}$$

The entropy of the liquid in equilibrium with the vapor is equal to the $S°$ for the gas calculated in a) at 1 Bar plus the value of ΔS calculated for expanding the gas from 1 Bar to the vapor pressure minus the enthalpy of vaporization. We neglect the entropy change in compressing the liquid from the vapor pressure to 1 Bar. Then

$$S_m^°(\text{liq}) = S_m^°(g) + 33.27 \text{J mol}^{-1} \ \text{K}^{-1} - \Delta S_m(\text{vap})$$

$$= (321.37 + 33.27 - 131.3) \ \text{J mol}^{-1} \ \text{K}^{-1}$$

$$= 223.3 \ \text{J mol}^{-1} \ \text{K}^{-1}$$

NBS tables give 223.0 J mol^{-1} K^{-1}. The agreement is good. Our calculation neglects the value of ΔS_m for the compression of the liquid from 1828 Pa to 0.1 MPa to reach the standard state. This calculation would require knowledge of $(\partial V_m/\partial T)_P$ for the liquid. If Exercise 10-15 is any guide, this correction would be too small to account for our discrepancy. By the rules of Parks and Huffman,

$$S_m^{\circ}(\text{liq}) = [104.65 + (2)(32.2) + (2)(38)] \text{ J mol}^{-1} \text{ K}^{-1}$$

$$= 245 \text{ J mol}^{-1} \text{ K}^{-1}$$

which is further from the NBS value than the value we calculated. Using more than one means of estimation is always advisable.

f) Pitzer's third law value is 223.3 J mol^{-1} K^{-1}, which is also close to the NBS value.

4.   a)

$$NC\text{-}CH_2\text{-}CH\text{=}CH\text{-}CH_2\text{-}CN$$

$$\Delta_f H_m^{\circ}$$

Base group, methane	- 74.85 kJ mol^{-1}
5 Primary CH$_3$ groups	5× - 9.83 kJ mol^{-1}
→CH$_3$-CH$_2$-CH$_2$-CH$_2$-CH$_2$-CH$_3$	
2 CN groups	2× 172.55 kJ mol^{-1}
2 Type 2	4× - 12.9 kJ mol^{-1}
1 2=2 double bond	116.40 kJ mol^{-1}
SUM	285.90 kJ mol^{-1}

$$S_m{}^\circ$$

Base group, methane	186.19 J mol^{-1} K^{-1}
5 Primary CH$_3$ groups	5× 43.30 J mol^{-1} K^{-1}
2 CN groups	2× 6.69 J mol^{-1} K^{-1}
2 Type 2	4× 2.3 J mol^{-1} K^{-1}
1 2=2 double bond	- 8.85 J mol^{-1} K^{-1}
SUM	416.42 J mol^{-1} K^{-1}

b) $C_4H_6 + (CN)_2 = C_4H_6(CN)_2$

              (1)   (2)   (3)

$$\Delta H_m^\circ = \Delta_f H_m^\circ(3) - \Delta_f H_m^\circ(1) - \Delta_f H_m^\circ(2)$$

$$= (285.90 - 111.914 - 300.495) \text{ kJ mol}^{-1} = -126.51 \text{ kJ mol}^{-1}$$

$$\Delta S_m^\circ = S_m^\circ(3) - S_m^\circ(1) - S_m^\circ(2)$$

$$= (416.42 - 277.90 - 241.17) \text{ J mol}^{-1} \text{ K}^{-1} = -102.65 \text{ J mol}^{-1} \text{ K}^{-1}$$

$$\Delta G_m^\circ = \Delta H_m^\circ - T\Delta S_m^\circ$$

$$= -126.51 \text{ kJ mol}^{-1} - (298.15K)(-102.65 \text{ J mol}^{-1} \text{ K}^{-1}) = -96.03 \text{ kJ mol}^{-1}$$

At 298.15 K,

$$K = e^{-\dfrac{\Delta G_m^\circ}{RT}}$$

$$= e^{-\dfrac{-96.03 \text{ kJ mol}^{-1}}{(8.3145 \text{ J mol}^{-1} \text{ K}^{-1})(298.15 \text{ K})}}$$

$$= 6.66 \times 10^{16}$$

This result suggests that the reaction is thermodynamically feasible.

Chapter 23

Practical Mathematical Techniques

2.   The table for numerical integration is shown in Table 23-1, as obtained from a spreadsheet.

Table 23-1.

T/K	C_{Pm}	C_{Pm}	$C_{Pm}\,dT$	$\Sigma C_{Pm}\,dT$
10	1.979			
15	6.125	4.052	20.260	
20	11.866	8.996	44.978	65.238
25	18.405	15.136	75.678	120.655
30	25.163	21.784	108.920	184.598
35	31.89	28.527	142.633	251.553
40	38.221	35.056	175.278	317.910
45	44.279	41.250	206.250	381.528
50	50.024	47.152	235.758	442.008
60	60.509	55.267	552.665	788.423
70	69.848	65.179	651.785	1204.450
80	78.287	74.068	740.675	1392.460
90	86.048	82.1675	821.675	3776.553

T/K	C_{Pm}	C_{Pm}	$C_{Pm}\,dT$	$\Sigma C_{Pm}\,dT$
100	92.772	89.410	894.100	1715.775
110	99.161	95.967	959.665	1853.765
120	105.286	102.224	1022.235	1981.900
130	111.156	108.221	1082.210	2104.445
140	116.922	114.039	1140.390	2222.600
150	122.784	119.853	1198.530	2338.920
160	129.131	125.958	1259.575	2458.105
170	136.394	132.763	1327.625	2587.200
180	144.499	140.447	1404.465	2732.090
182.55	146.595	145.547	371.145	1775.610

A graph of $\int_0^T C_{Pm}\,dT$ against T is shown in Figure 23-1, where the last column in the table is

taken equal to the integral from 15 K to each value of T. In order to obtain the lower limit of 0

K, we need to use the Debye equation to calculate $\int_0^{15\ K} C_{Pm}\,dT$.

$$C_{Pm} = aT^3$$

$$a = \frac{6.125 \text{ J mol}^{-1} \text{ K}^{-1}}{(15 \text{ K})^3} = 1.815 \times 10^{-3} \text{ J mol}^{-1} \text{ K}^{-4}$$

$$\int_0^{15 \text{ K}} C_{Pm} dT = \int_0^{10 \text{ K}} aT^3 dT = \frac{aT^4}{4} \Big|_0^{10 \text{ K}}$$

$$= \frac{1.815 \times 10^{-3} \text{ J mol}^{-1} \text{ K}^{-4}}{4} [(15 \text{ K})^4 - (0 \text{ K})^4]$$

$$= 22.97 \text{ J mol}^{-1} \text{ K}^{-1}$$

When 22.97 J mol^{-1} K^{-1} is added to each value in the last column of Table 32-1, we obtain the desired integral. In order to illustrate the region in which the Debye equation is needed, we will need to show the first few heat capacity data on an expanded scale, as shown in Figure 23-2.

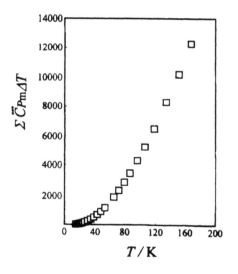

Figure 23-1. A plot of $\Sigma C_{Pm} \Delta T$ against T.

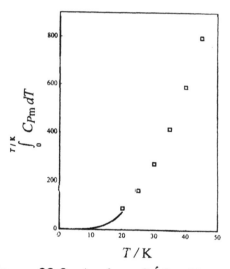

Figure 23-2. A plot of $\int C_{Pm} dT$ against T over a limited range of T and the Debye curve.